T0239073

SpringerBriefs on Cyber Security Systems and Networks

The series aims to develop and disseminate an understanding of innovations, paradigms, techniques, and technologies in the contexts of cyber security systems and networks related research and studies.

It publishes thorough and cohesive overviews of state-of-the-art topics in cyber security, as well as sophisticated techniques, original research presentations and in-depth case studies in cyber systems and networks. The series also provides a single point of coverage of advanced and timely emerging topics as well as a forum for core concepts that may not have reached a level of maturity to warrant a comprehensive textbook.

It addresses security, privacy, availability, and dependability issues for cyber systems and networks, and welcomes emerging technologies, such as artificial intelligence, cloud computing, cyber physical systems, and big data analytics related to cyber security research. The mainly focuses on the following research topics:

Fundamentals and theories

- Cryptography for cyber security
- Theories of cyber security
- Provable security

Cyber Systems and Networks

- Cyber systems security
- Network security
- Security services
- Social networks security and privacy
- Cyber attacks and defense
- Data-driven cyber security
- Trusted computing and systems

Applications and others

- Hardware and device security
- Cyber application security
- Human and social aspects of cyber security

More information about this series at https://link.springer.com/bookseries/15797

Jin Li · Ping Li · Zheli Liu · Xiaofeng Chen ·
Tong Li

Privacy-Preserving Machine Learning

Jin Li
School of Computer Science and Cyber
Engineering, Institute of Artificial
Intelligence and Blockchain
Guangzhou University
Guangzhou, Guangdong, China

Zheli Liu
College of Cyber Science and College
of Computer Science
Nankai University
Tianjin, China

Tong Li
College of Cyber Science and College
of Computer Science
Nankai University
Tianjin, China

Ping Li
School of Computer Science
South China Normal University
Guangzhou, Guangdong, China

Xiaofeng Chen
State Key Laboratory of Integrated Service
Network
Xidian University
Xi'an, China

ISSN 2522-5561　　　　　　ISSN 2522-557X　(electronic)
SpringerBriefs on Cyber Security Systems and Networks
ISBN 978-981-16-9138-6　　　ISBN 978-981-16-9139-3　(eBook)
https://doi.org/10.1007/978-981-16-9139-3

This Springer imprint is published by the registered company Springer Nature Singapore Pte Ltd.
The registered company address is: 152 Beach Road, #21-01/04 Gateway East, Singapore 189721, Singapore

Preface

As an implementation methodology of the artificial intelligence, machine learning techniques have reported impressive performance in a variety of application domains, such as risk assessment, medical predictions, and face recognition. Due to critical security concerns, how to protect data privacy in machine learning tasks has become an important and realistic issue spanning multiple disciplines. An ever-increasing number of researches have started proposing countermeasures to mitigate the threats of privacy leaks.

After motivating and discussing the meaning of privacy-preserving techniques, this book is devoted to provide a thorough overview of the evolution of privacy-preserving machine learning schemes over the last 10 years. We report these works according to different learning tasks.

In a learning task, a natural question is how the participants take the advantage of cooperative learning (Chap. 2) on the joint dataset of all participants' data while keeping their data private. Alternatively, besides the basic security requirements, the participants could suffer some bottlenecks on resources of computation, communication, and storage. Thus, they can outsource their computation workloads to cloud servers and enjoy the unlimited computation resources in a secure outsourced learning (Chap. 3) manner.

Massive data collection required for large-scale deep learning not only presents obvious privacy issues but also introduces the problems of efficiency and parallelization. The framework of distributed federated learning (Chap. 4) is necessary, by which the optimization algorithms used in deep learning can be parallelized and executed asynchronously. Moreover, to prevent learning results exposing private individual information in the dataset, the federated learning algorithm is supposed to achieve the differential privacy (Chap. 5) which is a strong standard for privacy guarantees for random algorithms on aggregate datasets.

Nowadays, machine learning classification is used for many data-driven applications. So, it is important to consider secure inference techniques (Chap. 6), in which the data and the classifier remain confidential when a user queries a classifier not owned by him/her. In Chap. 7, we turn to a concrete application, *i.e.*, privacy-preserving image processing.

This book is meant as a thorough introduction to the problems and techniques but is not intended to be an exhaustive survey. We can cover only a small portion of works of privacy-preserving machine learning.

Guangzhou, China   Jin Li
Guangzhou, China   Ping Li
Tianjin, China   Zheli Liu
Xi'an, China   Xiaofeng Chen
Tianjin, China   Tong Li

Contents

Chapter 1
Introduction

1.1 What Is Machine Learning?

Normally, we have many ways to explain machine learning. Before giving the definition, we would like to illustrate why we need machine learning. To give a solution of a computational problem, such as sorting, we need to carried out an algorithm to transform the problem's input to an output. But sometimes, we may not have exactly algorithm to give a solutions for some tasks. For example, our task is to select spam emails from a massive undetected emails. If there is an algorithm can help us finish this task, we only know that its input is the feature of an email and its output can indicate whether this email is spam or not. However, we do not know how to design such an algorithm that makes a transforming mapping from an input to an output. Let's see what we can do. Well, we can collect a large volume of emails and label them according to our knowledge of the spam. It is interesting that we wish to have a machine effective enough to produce automatically programs for this task, while we can only provide these example data. If only there had been such a machine driven by these data!

Fortunately, where there is a will, there is a way. With advances of the data and computer technology, we currently have the capability to store and process large amounts of data locally as well as over a remote server or networks. Collecting reliable data does not seem to be difficult for us since most data acquisition devices are embedded and omnipresent now. Intuitively, the collected data can become useful to guide our decisions only when it is extracted into knowledge that we can make use of. There should be a process that explains the data observed, though the details of the process underlying the data could be unknown. There may be certain patterns in the data, as there may be some spam-level properties in an email. Although identifying patterns completely may not be possible, a good and useful approximation is supposed to construct, which can account for some part of the data. If the future statement will not be much different from the past experiences from the sample data, evaluations can also be expected to be right in the future. We need a methodology to make full use of such experiences. This is why we need machine learning.

J. Li et al., *Privacy-Preserving Machine Learning*, SpringerBriefs on Cyber Security Systems and Networks, https://doi.org/10.1007/978-981-16-9139-3_1

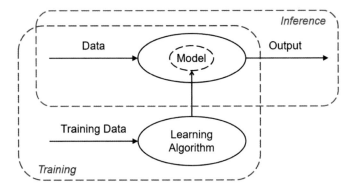

Fig. 1.1 Machine learning system

Machine learning is a methodology to study how to use experiences to improve the performance of a system on a specific set of tasks. Usually, the experiences exist in the form of data in computer systems. Therefore, one of the most important applications of machine learning methods is processing a large volume of data to construct a predictive model with valuable use, such as high accuracy evaluations. Moreover, the machine learning task is a part of artificial intelligence rather than just a problem over datasets. To make a system intelligent, it should have the ability to adaptively learn in various environments, such that its designer need not provide solutions for all possible situations. Machine learning also finds applications in diverse fields such as modelling, time series analysis, pattern recognition, and signal processing. Of course, it can be applied for identifying spam emails in the example above. Every day we receive emails sent from mailboxes everywhere, despite differences in form, title, content, and so on. After reading a mail, we can easily determine whether it is a spam or not and thus label it. But we are unable to explain how we do it since our expertises can be not be explained by writing a computer program. Alternatively, it is acknowledge that spam contents are not just a random collection of characters. By analyzing sample emails, each of which has its features, a learning algorithm is supposed to capture the specific pattern of spams and then check whether this pattern is matched in a new given email.

Figure 1.1 shows the stages of a machine learning system. The target of machine learning is to optimize the performance criterion of a system using past experiences which are presented as example data. In more details, a model needed to be learned is defined by some parameters, and the learning task is to execute an algorithm to minimize differences between these parameters and a "ideal" model's parameters using the example data. This model can be used to make predictions in the future. Normally, machine learning uses the statistics theory in building mathematical models for making inference from new examples. The computer science plays a role here in twofold. On one hand, in training, it is necessary to give efficient algorithms to solve the optimization problem. On the other hand, once a model has been trained, its algorithmic solution for future inferences also needs to be efficient. In applications in the real world, the efficiency of training or inference algorithms may be as important as the predictive accuracy.

In this book, our focus is mainly on privacy-preserving techniques for supervised machine learning which requires labelled training data, but it does not means unsupervised learning is unimportant. To illustrate a concrete task in Fig. 1.1, we still take "spam emails" as the example. We assume that there are a set of sample of emails and a group of people that can manually select these mails. The emails are observed and labelled, such that spam mails are labelled as positive samples while the other are labelled as negative samples. The learning task here is to find a "notion" that is defined by all positive samples and none of the negative samples. To do this, an evaluation can be made as follows. Given a new mail has not been seen before, we can decide whether it is a spam mail or not by checking with each part learned. This task may be built by a mail website, and the aim may be to determine and block spam mails. The features that separate a spam mail from other mails could be title, content, and other part. These attributes are the inputs to the trained model $\mathbf{W}$ when conducting inferences.

Inference. The inference of a machine learning model can be seen as parametric functions $C_{\mathbf{W}}(\mathbf{x})$ based on the model $\mathbf{W}$. Such a function can be used for predicting the input $\mathbf{x}$ that is often represented as a vector of feature values. The model $\mathbf{W}$ for the spam detection would predict whether a new mail $\mathbf{x}$ is more likely to be a spam or legal mail. Taking the neural network as an example, the prediction result is decided by neuron values at the output layer of $\mathbf{W}$. In the most common case for classification, the result is a soft-max vector assigning a probability for each class, which indicates how likely the input is to belong to that class.

Training. The training is a process of adjustments applied to *connection weights* between neurons according to training data. In another word, after collecting and pre-processing training data, we improve the performance of $C_{\mathbf{W}}(\mathbf{x})$ by analyzing the training data and iteratively seeking the values of $\mathbf{W}$. For "spam emails", the training is supervised, so that the training dataset should be represented by a set of input-output examples (i.e., $\mathbf{x}$-$\mathbf{t}$). Thus, the model $\mathbf{W}$ are adjusted to reduce the gap between the prediction $C_{\mathbf{W}}(\mathbf{x})$ and the desired output $\mathbf{t}$ of each instance $\mathbf{x}$ in the dataset. The performance of a trained model $\mathbf{W}$ will be validated on a testing dataset disjoint from the training dataset. This validation can be seen as a series of "special" *inferences*. Normally, the trainer interests in achieve a high accuracy of $\mathbf{W}$ on a test dataset rather than the training dataset.

We can see that the spam email recognition above only involves positive samples and negative samples. So, it constitutes a binary classification task, which is easily the most common task of machine learning. Sometimes, we consider problems other than the two-class classification in applications.

Multi-Class Classification. We can extend the recognition to a multi-class classification task as an example. We may want to distinguish different kinds of received mails. The multi-class classification could be approached as a combination of several binary classification tasks, but some potentially useful information may get lost this way. For this reason, it is often beneficial to view multi-class classification as a machine learning task in its own right.

Regression. When needing to meet some requirements, we can also predict a real number rather than use the notion of discrete classes. In the example of "spam emails", the system may requires to evaluate the urgency of an email on a sliding scale. We call this task regression which essentially involves learning a real-valued function from training examples labelled with true function values. We still use the example above to explain it. The training set are constructed from a number of emails labelled with a score on a scale of 0 (i.e., ignore) to 10 (i.e., red flag). This typically works by choosing a class of functions and constructing a function which can minimize the difference between predicted values and "true" function values. In a regression task it is necessary to find a way to express a model's confidence in its real-valued predictions.

Clustering. Both classification and regression assume the availability of a training set of examples labelled with true classes or function values. Here we give an example of unsupervised learning that is quite distinct from supervised learning. Maybe providing the true labels for a data set is often expensive. Thus, we can only learn to distinguish spam emails without a labelled training set. The task of grouping data without prior information on the groups is called clustering which typically works by assessing the similarity between instances and putting similar instances in the same cluster.

As mentioned above, all these machine learning problems don't have a exactly correct answer. This is different from many other familiar problems in computer sciences, even though there may be many algorithms of achieving the correct result. To compare the performance of these algorithms, it would be in terms of how fast they are rather than how accurate they are. Things are different in machine learning. Assume that the perfect spam e-mail filter does not exist. If it existed, spammers would immediately reverse it to find out ways to trick the spam filter into thinking a spam e-mail is actually ham. Considering data may be mislabelled or features may contain errors, it would be detrimental to try too hard to find a model that correctly classifies the training data. In some cases the features used to describe the data only give an indication of what their class might be, but not contain enough Information to predict the class perfectly. In all these procedures, we can see that enough and useful training data are very important. This is the starting point for our discussion on privacy-preserving issues of machine learning.

1.2 Why Machine Learning Needs Privacy-Preserving Manner

Let's review what we should do if we want to train a machine learning model. Firstly, we collect raw data and perform the pro-processing on them for training. Then, taking a set of data, we run a learning algorithm to output a trained model. Maybe we need some test data for evaluating the performance of the model. We can see that if all these procedures are conducted locally, there is no need for protecting data privacy.

In another word, the local environment is totally trusted, so that we do not consider any risk of privacy leakage even the data are sensitive.

However, the local learning is not everything. Normally, to make a machine learning model perform better in a target task, we need massive relevant data that can reflect the population of data as much as possible. The data in the real world are constantly growing, and thus collecting enough data sometimes can hardly be done by oneself. Furthermore, some data have characteristics which are restricted by regions, authorities, or expertises. The model trained on someone's dataset will fit its local characteristics rather than capture the population's characteristics. Although the generalization can avoid the overfitting for which the correct output is not given in the training data, it still cannot overcome that most of the data is missing. These inherent defects motivate us to cooperate with other trainers for improving a learning task. That means all of trainers should contribute their training data for the cooperative learning and set up the task in an open environment.

In past decade, with the rapid development of network and computation techniques, the cooperative learning has become reasonable and easily be deployed. As a new deployment and delivery model of computing resources in the open environment, cloud computing enables convenient network access to a virtualized pool of remote resources. Besides data sharing, users can acquire more related computation services from a powerful remote server. Obviously, the open environment benefits the data collection and sharing from unlimited resources as well as the efficient learning among multiple participants.

The opportunities, however, always come with challenges. Under an open environment, such as the cloud environment, we cannot ensure that all the entities are fully trusted. In another word, the resources and services of a machine learning task are open for public use and communication is performed over a untrusted network. The participants of a machine learning task, including a public cloud server perhaps, are not fully trusted by each other. Therefore, although the benefits under open environments are tremendous, serious security risks and privacy challenges are raised under the untrusted environments. In more details, when learning participants contribute their data to others for training, evaluation, or other use, they will lose tight control of the data as in their local learning tasks. Curious service providers, participants, and eavesdroppers may deliberately access the data not belongs to them and obtain the sensitive information. Moreover, data corruption or tampering could also happen during training due to the attacks initiated by malicious adversaries. Generally, when learning participants cooperatively training a model, privacy breaches often occur due to undesirable interference from internal and external adversaries. For example, when training a model for medical diagnoses, a participant may learn any patient's health record by observing medical data in training dataset. Thus, it can be seen that procedures in a machine learning task are intrinsically not secure from the viewpoint of all participants. If there is no effective security and privacy protection guarantee, we are hard to believe that participants will delegate others to use their data for learning in a desirable way. That's why we introduce privacy-preserving manners in machine learning, which is to alleviate these privacy concerns.

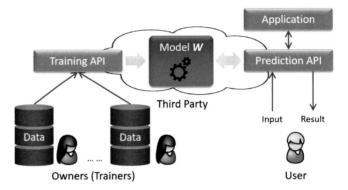

Fig. 1.2 The system model of learning task

As shown in Fig. 1.2, the learning task in the open environment (Fig. 1.2) is different from the one running locally (Fig. 1.2). As mentioned above, there are more than one trainers participate in this task and maybe some third-party entities. The trainers share their local datasets with others for cooperative learning over the sum aggregate of these datasets. The third party, is sometimes a remote server with the powerful data calculate capability, is optional to offer services such as data computation and data distributing for solving bottlenecks of large-scale learning tasks. After training, the well-trained model may be used for providing a service such as time series analysis, pattern recognition, and signal processing.

Due to the untrusted entities and the external environment, security risks are almost everywhere and threaten the data privacy in the learning task. Here, we use the medical big data as an example to illustrate them. To build a tool to predict the health condition of a patient, we consider to train a machine learning model which takes the patient's health record as input and outputs the evaluation result. A group of medical researchers make an agreement with each other to build such a model on their jointly shared records across a large population of patients. However, in the open environment, (e.g., the cloud environment), these trainers do not totally trusted for each other. Moreover, health records frequently contain sensitive information such that the privacy of patients must be protected according to some privacy rules such as Health Insurance Portability and Accountability Act (HIPAA) which establishes the regulations for the use and disclosure of Protected Health Information. Therefore, we have to solve a security problem: how to cooperative train the machine learning model while keeping each trainer's dataset private, even if communications, remote storages, and computations are left open to attacks. In another word, the challenge here is to give techniques that protect the privacy of each trainer while enabling the learning task above.

1.3 Security Threats

There are quite a number of security threats associated with machine learning in the open environment. From the viewpoint of participants of the learning task, two types of adversaries are considered to pose security threats to data privacy. One is the internal attacker and the other is the external attacker. Obviously, compared to an external attackers (e.g., a hacker), an internal participant has more background knowledge about the training dataset, e.g., a part of the dataset or intermediate training results. That is, the internal attacker is more powerful than the external attacker. Therefore, we only consider internal attackers in the rest of the book. We can model these adversaries from three dimensions as follows.

Target. The target of attacks refers to which sensitive information the adversary wants to reveal. As mentioned above, a training dataset contains sensitive information of its owner. That is why we require the training procedures running in a privacy-preserving way. Sometimes, the training result, i.e., a well-trained machine learning model, also contains sensitive sensitive and valuable information, since it is trainers' labour with intellectual property rights. For example, a search service for pictures is equipped with an image recognition model which is built by complicated endeavours on thousands of training samples. The detail contents of such a model, e.g., parameters and hyper-parameters, are valuable, whose leakage will harm the economic interest of the service provider. Therefore, the model can be also seen as the private information which the adversary targets. Moreover, if we deploy this service on the public environment, revealing the model is not the only goal of adversaries. Maybe, the querying data which provided by service users for classification or recognition contain sensitive information. Well, adversaries will pay more attentions on the privacy of users. To sum up, attack targets of adversaries include training data, machine learning models, and querying data. Sometimes, the adversary is interested in a sort of data, e.g., data with a special label, such that it aims at revealing these data rather than the whole dataset. We call its attack *specific*, while the attack targeting on the whole dataset is *general*.

Knowledge. The knowledge refers to the adversary's understanding of its attacking target. For a dataset, its background, i.e., population, statistical information, a part of the dataset, can help to inference the targeted content of the dataset. Especially, a well-trained model is also background knowledge of the dataset. This is because an useful model will reveal more about the input to which it is applied than would have been known about this input without applying the model. Obviously, it is the easiest to consider our privacy-preserving manner when the adversary knows nothing about the dataset. For a model in classification services, an adversary which tries to reveal its content is normally allowed to get access to it several times. Besides a set of "input-output" pairs, the adversary may know the model's structure, hyperparameters, or nothing.

Capability. Generally, an attacker is either *passive* (e.g., honest-but-curious) or *active*. A passive adversary can honestly execute the computation protocols and faithfully provide datasets, but it is curious about the data not belongs to it. Thus, it desires to learn the sensitive information (e.g., identity, preferences, and habits) in the data stored or processed in the open environment. Normally, this adversary may be an eavesdropper, a learning participant, or a cloud server. Different from the passive adversary, an active one may utilize a variety of attack techniques to attempt to get unauthorized access to the data. This adversary may intentionally tamper with or forge the data processed during learning tasks for its unfavored motivations, which will help it reveal private information more accurately. That is, it may maliciously infer the privacy out of curiosity and not ensure the correctness of learning results.

The security is measured with respect the adversarial targets and capabilities, such that a privacy-preserving machine learning scheme is designed to defend against the specific adversaries. Then, we identify the security threats of machine learning tasks by when an adversary will attempt to reveal the private information under attack. As shown in Table 1.1, we discuss the attacks as they relate to inference and training phases.

- **Eavesdropping**. The eavesdropping is a common attack for breaking data privacy, which is usually initiated by passive adversaries. An external eavesdropper can capture some messages in the protocol during a machine learning task. An internal participant may be more powerful than the external one. For example, some third-party servers can be introduced in the learning task to solve computational bottlenecks. Besides the captured messages, such servers temporarily store several intermediate results which may contain some underlying information about the training dataset. They can honestly execute the learning protocols and faithfully provide data services, but they are curious about the processed data. They desire to learn the information of the dataset with the help of their knowledge.
- **Data Reconstruction**. The data reconstruction attack is to reconstruct training data not belongs to the attacker during a cooperative learning task. We consider the following case. We have some users which store local datasets of private information on their respective devices and try to cooperate to build a global model. A malicious user simply runs the collaborative learning algorithm and reconstructs sensitive information stored on the victim's device. It is able to influence the learning process and deceive the victim into releasing more detailed information.
- **Member Inference**. The member inference attack is to determine whether a given record was used as part of a given machine learning model's training dataset or not. Such an attack usually occurs when a service provider offers "machine learning as a service". The adversary gets access to the model and makes some black-box queries that return the model's output on a purposely chosen input. Then, it solves the member inference problem through the prediction outputs of machine learning models.
- **Model Stealing**. The model stealing attack is to duplicate the functionality of a machine learning model not belongs to the adversary. The adversary initiates the attack with black-box access but no prior knowledge of the model's parameters, when using the service based on this model. The service may accept partial feature vectors as inputs and include confidence values with predictions. Therefore, effi-

Table 1.1 Threats in machine learning tasks

Phase/target	Training data	Model
Training	Eavesdropping/Data reconstruction	–
Inference	Member inference	Model stealing

cient attacks that extract target some models with near-perfect fidelity for popular model classes including logistic regression, neural networks, and decision trees.

In the rest of this book, we will focus on privacy issues of machine learning and introduce some typical solutions for mitigating or getting rid of the threats.

1.4 Bibliographic Notes

The literature Mitchell (1997) is an excellent specialized textbook of machine learning. Literatures Alpaydin (2004), Flach (2012), Hart et al. (2000) are valuable monographs which can help readers to master machine learning in a faster and easily way. Hastie et al. (2009) is an advance textbook of machine learning. Anzai (2012) is also of great reference value for those who prefer Bayesian decision, while Shalev-Shwartz (2014) is suitable for those who prefer theory. Witten and Frank (2002) is an introduction of WEKA which helps beginners to quickly learn commonly used machine learning algorithms through WEKA. Many early literatures are still worthy of attention today. The third volume of the book "Artificial Intelligence Manual" Cohen and Feigenbaum (2014) mainly discussed machine learning, which is an important reference in the early stage of machine learning. Dietterich (1997) reviews and prospects for the development of machine learning. A recent popular technique named "transfer learning" provides the great development of statistical learning, which considers multiple participants and datasets in learning tasks. Today, an upgraded version of deep learning has been widely adopted in many field of Artificial Intelligence.

With the rapid development of cloud computing, more and more enterprises/ individuals are starting to develop solutions for large-scale computational tasks with the help of cloud services. Moreover, recent researches have made advances in addressing privacy concerns in cloud computing for computationally intensive applications. New encryption primitives (i.e., searchable encryption Boneh et al. (2004), Wang et al. (2010) and homomorphic encryption Gentry (2009), Smart and Vercauteren (2010)) have been proposed to enable secure data search and data computation over an untrusted server. Some solutions combined fully homomorphic encryption (FHE) Gentry (2009) with the evaluation of Yaos garbled circuits (GCs) Yao (1982) to achieve both data confidentiality and result verifiability. The survey Tang et al. (2016) presented the data privacy issues and privacy-preseving techniques under the untrusted environment.

There have been many researches that address privacy problems in multi-party machine learning tasks, and most of these works focus on defending against the adversaries acting as eavesdroppers. They cover techniques such as random decision trees Vaidya et al. (2014), Naive Bayes classification Vaidya et al. (2008, 2013), k-means clustering Jagannathan and Wright (2005), and linear programming Vaidya (2009). Some of the works give mechanisms for protecting the privacy of sample data during training a classifier over the data. The others concern protecting a trained classifier when using it to classify instances Barni et al. (2011), Bost et al. (2014).

For the privacy-preserving neural network learning, earlier works Schlitter (2008), Chen and Zhong (2009), Bansal et al. (2011), Samet and Miri (2012) mainly paid attention on how to protect a training dataset partitioned into different forms. They were equivalent to each trainer secretly owning either a few records (a horizontally partitioned sub-set) or a few vector components (a vertically partitioned sub-set) of the set. Since the global model is updated according to the training dataset, trainers will communicate with each other several times for processing protected data. Switching intermediate results forth and back significantly introduces overhead to trainers (data owners). Furthermore, some schemes (e.g., Li et al. 2017; Xie et al. 2014) adopted a high-cost cryptographic tools, such as a FHE scheme or a garble circuit Yao (1982) based Multi-Party Computation (MPC) protocol Lindell (2016), to protect the data privacy while ensuring the data usability, which makes the schemes far from practical applications. Yuan et al. (2014) proposed a server-aid learning scheme, where trainers that hold arbitrarily partitioning datasets perform training with the help of a cloud server. Mohassel et al. (2017) proposed a privacy-preserving machine learning scheme that oursources training tasks using a secure two-party-computation technique. Mohassel et al. (2018) speeded up outsourcing learning schemes by using a three-party-computation architecture, in which three servers jointly perform training and evaluate the trained model. Graepel et al. (2012), Nikolaenko et al. (2013, 2013) worked for the mixture of privacy and communication issues in the cooperative training of other machine learning classifiers. Graepel et al. (2012) proposed a "somewhat" general learning scheme based on additive homomorphic encryption. In some cooperative privacy-preserving training tasks, a global model can be computed by aggregating local parameters. Shokri et al. (2015) proposed a privacy-preserving federated learning system that enables data owners to parallelizedly train a neural network classifier model. Abadi et al. (2016) proposed a deep learning scheme with differential privacy Dwork and Roth (2014). In this work, data owners should share sensitivity Dwork and Roth (2014) of the whole dataset and aggregately add noises on the gradients in each iteration, so that a privacy mechanism can be implemented on the final model. Ohrimenko et al. (2016) gave a solution for the data-oblivious multi-party machine learning by using trusted SGX-processors. To reduce the communication overhead and enhance the robustness for dropping out in federated learning systems, Bonawitz et al. (2017) proposed a secure aggregation protocol.

Besides the eavesdroppers, the attacks initiated by other types of adversaries are considered by researchers. Hitaj et al. (2017) shows the leakage of sensitive information from the collaborative deep learning task if there is any active adversary.

The scheme allows the adversary to train a Generative Adversarial Network (GAN) that reconstructs specific samples of the targeted training set not owned by it. Shokri et al. (2017) study how machine learning models leak information about a specific training data record. They propose an attack approach named "membership inference attack". Given a data record and black-box access to a model, an adversary can initiate the attack determine if this record was in the model's training dataset. Tramer et al. (2016) propose attacks that steal the parameters of target machine models with near-perfect fidelity for popular models, such as logistic regression, neural networks, and decision trees.

References

Abadi M, Chu A, Goodfellow I, McMahan HB, Mironov I, Talwar K, Zhang K (2016) Deep learning with differential privacy. IDn: Proceedings of the 2016 ACM SIGSAC conference on computer and communications security. ACM, pp 308–318

Alpaydin E (2004) Introduction to machine learning (Adaptive computation and machine learning). MIT Press

Anzai Y (2012) Pattern recognition and machine learning. Elsevier

Bansal A, Chen T, Zhong S (2011) Privacy preserving back-propagation neural network learning over arbitrarily partitioned data. Neural Comput Appl 20(1):143–150

Barni M, Failla P, Lazzeretti R, Sadeghi A-R, Schneider T (2011) Privacy-preserving ECG classification with branching programs and neural networks. IEEE Trans Inf Forensics Sec 6(2):452–468

Bonawitz K, Ivanov V, Kreuter B, Marcedone A, McMahan HB, Patel S, Ramage D, Segal A, Seth K (2017) Practical secure aggregation for privacy-preserving machine learning. In: Proceedings of the 2017 ACM SIGSAC conference on computer and communications security. ACM, pp 1175–1191

Boneh D, Di Crescenzo G, Ostrovsky R, Persiano G (2004) Public key encryption with keyword search. In: Cachin C, Camenisch JL (eds), Advances in cryptology - EUROCRYPT, pp 506–522

Bost R, Popa RA, Tu S, Goldwasser S (2014) Machine learning classification over encrypted data., IACR Cryptology ePrint Archive 2014 331

Chen T, Zhong S (2009) Privacy-preserving backpropagation neural network learning. IEEE Trans Neural Netw 20(10):1554–1564

Cohen PR, Feigenbaum EA (2014) The handbook of artificial intelligence: Volume 3, vol 3. Butterworth-Heinemann

Dietterich TG (1997) Machine-learning research. AI Mag 18(4):97

Dwork C, Roth A et al (2014) The algorithmic foundations of differential privacy, Foundations and Trends®. Theoret Comput Sci 9(3–4):211–407

Flach P (2012) Machine learning. The art and science of algorithms that make sense of data. Cambridge University Press

Gentry C (2009) A fully homomorphic encryption scheme, PhD thesis, Stanford University

Graepel T, Lauter K, Naehrig M (2012) Ml confidential: machine learning on encrypted data. In: International conference on information security and cryptology, Springer, pp 1–21

Hart PE, Stork DG, Duda RO (2000) Pattern classification. Wiley Hoboken

Hastie T, Tibshirani R, Friedman J (2009) An introduction to statistical learning

Hitaj B, Ateniese G, Perez-Cruz F (2017) Deep models under the gan: information leakage from collaborative deep learning. In: Proceedings of the 2017 ACM SIGSAC conference on computer and communications security, pp 603–618

Jagannathan G, Wright RN (2005) Privacy-preserving distributed k-means clustering over arbitrarily partitioned data. In: Proceedings of the eleventh ACM SIGKDD international conference on Knowledge discovery in data mining. ACM, pp 593–599

Li P, Li J, Huang Z, Li T, Gao C-Z, Yiu S-M, Chen K (2017) Multi-key privacy-preserving deep learning in cloud computing. Futur Gener Comput Syst 74:76–85

Lindell Y (2016) Fast cut-and-choose-based protocols for malicious and covert adversaries. J Cryptol 29(2):456–490

Mitchell T (1997) Machine learning

Mohassel P, Rindal P (2018) Aby 3: a mixed protocol framework for machine learning. In: Proceedings of the 2018 ACM SIGSAC conference on computer and communications security. ACM, pp 35–52

Nikolaenko V, Ioannidis S, Weinsberg U, Joye M, Taft N, Boneh D (2013) Privacy-preserving matrix factorization. In: Proceedings of the 2013 ACM SIGSAC conference on computer & communications security. ACM, pp 801–812

Nikolaenko V, Weinsberg U, Ioannidis S, Joye M, Boneh D, Taft N (2013) Privacy-preserving ridge regression on hundreds of millions of records. In: 2013 IEEE symposium on security and privacy (SP). IEEE, pp 334–348

Ohrimenko O, Schuster F, Fournet C, Mehta C, Nowozin S, Vaswani K, Costa M (2016) Oblivious multi-party machine learning on trusted processors. In: USENIX security, vol 16, pp 619–636

p. Mohassel P, Zhang Y (2017) Secureml: a system for scalable privacy-preserving machine learning. In: IEEE Symposium on security and privacy (SP). IEEE, pp 19–38

Samet S, Miri A (2012) Privacy-preserving back-propagation and extreme learning machine algorithms. Data Knowl Eng 79:40–61

Schlitter N (2008) A protocol for privacy preserving neural network learning on horizontally partitioned data, PSD

S. Shalev-Shwartz, S. Ben-David, Understanding machine learning: From theory to algorithms, Cambridge university press (2014)

Shokri R, Shmatikov V (2015) Privacy-preserving deep learning, in: Proceedings of the 22nd ACM SIGSAC conference on computer and communications security. ACM, pp 1310–1321

Shokri R, Stronati M, Song C, Shmatikov V (2017) Membership inference attacks against machine learning models. In: IEEE symposium on security and privacy (SP). IEEE, pp 3–18

Smart NP, Vercauteren F (2010) Fully homomorphic encryption with relatively small key and ciphertext sizes. In: International workshop on public key cryptography. Springer, pp. 420–443

Tang J, Cui Y, Li Q, Ren K, Liu J, Buyya R (2016) Ensuring security and privacy preservation for cloud data services. ACM Comput Surv 49(1):1–39

Tramèr F, Zhang F, Juels A, Reiter MK, Ristenpart T (2016) Stealing machine learning models via prediction apis. In: 25th {USENIX} security symposium ({USENIX} Security 16), pp 601–618

Vaidya J (2009) Privacy-preserving linear programming. In Proceedings of the 2009 ACM symposium on applied computing. ACM, pp 2002–2007

Vaidya J, Kantarcıoğlu M, Clifton C (2008) Privacy-preserving naive bayes classification,. VLDB J Int J Very Large Data Bases 17(4):879–898

Vaidya J, Shafiq B, Fan W, Mehmood D, Lorenzi D (2014) A random decision tree framework for privacy-preserving data mining. IEEE Trans Dependable Secure Comput 11(5):399–411

Vaidya J, Shafiq B, Basu A, Hong Y (2013) Differentially private naive bayes classification. In: Proceedings of the 2013 IEEE/WIC/acm international joint conferences on web intelligence (WI) and intelligent agent technologies (IAT)-Volume 01. IEEE Computer Society, pp 571–576

Wang C, Cao N, Li J, Ren K, Lou W (2010) Secure ranked keyword search over encrypted cloud data. In: IEEE 30th international conference on distributed computing systems. IEEE, pp. 253–262

Witten IH, Frank E (2002) Data mining: practical machine learning tools and techniques with java implementations. ACM SIGMOD Rec 31(1):76–77

Xie P, Bilenko M, Finley T, Gilad-Bachrach R, Lauter K, Naehrig M (2014) Crypto-nets: Neural networks over encrypted data. arXiv:1412.6181

Yao AC (1982) Protocols for secure computations. In 23rd annual symposium on foundations of computer science, SFCS'08. IEEE, pp 160–164

Yuan J, Yu S (2014) Privacy preserving back-propagation neural network learning made practical with cloud computing. IEEE Trans Parallel Distrib Syst 25(1):212–221

Chapter 2
Secure Cooperative Learning in Early Years

2.1 An Overview of Neural Network

Artificial Neural Networks (ANN) has been motivated right from its inception by the recognition that the human brain, which is a highly complex, non-linear, and parallel-computing, works in an entirely different way from the conventional digital computer. Viewed as an adaptive machine, an ANN is a massively parallel distributed processor made up of simple processing units, which has a natural propensity for storing experiential knowledge and making it available for use. An ANN is able to learn and generalize from input data, so that its outputs is reasonable for "new" inputs not encountered during learning (training). The parallel-distributed structure and the learning ability make it possible for the ANN to find good approximate solutions to large-scale problems that are intractable. The ANN represents a technology that finds applications in diverse fields such as modelling, time series analysis, pattern recognition, and signal processing.

A *neuron* is a basic information-processing unit that is fundamental to operations of a neural network. The model of a neuron is shown in Fig. 2.1. In this model, the neuron connects with other n neurons each of which inputs a signal x_i via an income connection. A connection is characterized by a weight w_i of its own. After receiving the signals, the neuron will linearly combine the input signals and then compare the result to a threshold θ. Finally, an activation function $f(\cdot)$ limits the amplitude of the output of this neuron. The output is

$$y = f\left(\sum_{i=1}^{n} w_i x_i - \theta\right). \tag{2.1}$$

The sigmoid function is the most common form of activation functions used in the construction of neural networks. As shown in Fig. 2.2, the graph of the function is s-shaped. It is defined as a strictly increasing function that squashes an input value

J. Li et al., *Privacy-Preserving Machine Learning*, SpringerBriefs on Cyber Security Systems and Networks, https://doi.org/10.1007/978-981-16-9139-3_2

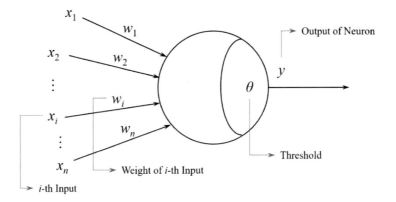

Fig. 2.1 Neuron model

Fig. 2.2 Sigmoid function

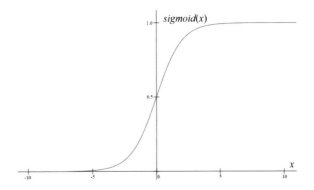

chosen from a large domain into a continuous range of values from 0 to 1. The function is defined by

$$sigmoid(x) = \frac{1}{1 + e^{-x}}. \qquad (2.2)$$

A neural network is composed of several connected neurons by a certain hierarchical structure. Figure 2.3 describes a neural network called the multilayer feed-forward network. We have an *input layer* of source nodes that projects onto one or more subsequent *hidden layers* of intermediate nodes that projects onto an *output layer* of neurons. Note that this network is strictly feed-forward and acyclic. The *neural network* that we talk about here is the "neural network model", and we use **W** to represent the model of a neural network.

Fig. 2.3 Multilayer
feed-forward network

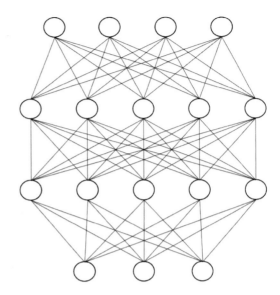

2.2   Back-Propagation Learning

In this chapter, we focus on the training of the multilayer perception. Multilayer feed-forward networks have been applied successfully to solve some difficult diverse problems by training them in a supervised manner. A popular method for such a training is the *Back-Propagation (BP) algorithm*, which is based on the error-correction learning rule.

   Given a training set $D = \{(\mathbf{x}^{(1)}, \mathbf{t}^{(1)}), ..., (\mathbf{x}^{(m)}, \mathbf{t}^{(m)})\}$ where $\mathbf{x}^{(l)} \in \mathbb{R}^a$ and $\mathbf{t}^{(l)} \in \mathbb{R}^c$, we take the learning process of a 3-layer neural network shown in Fig. 2.4 as an example. This network with *a-b-c* configuration contains one hidden layer. In another word, a, b, and c are the number of neuron nodes at the input layer, the hidden layer, and the output layer, respectively. The threshold of the jth node at the hidden layer is denoted as γ_j, while the threshold of the ith node at the output layer is denoted as θ_i. For the connections, $w_{jk}^{(h)}$ denotes the weight between input layer node k and hidden layer node j, while $w_{ij}^{(o)}$ denotes the weight between hidden layer node j and output layer node i. The input value of the jth node at the hidden layer is $\alpha_j = \sum_{k=1}^{a} w_{jk}^{(h)} x_k$, while the input value of the ith node at the output layer is $\beta_i = \sum_{j=1}^{b} w_{ij}^{(o)} h_j$. We set that all the neuron nodes at the hidden layer and the output layer use the sigmoid function $f(\cdot)$ as the activation function.

   Assuming that the input training instance is $(\mathbf{x}, \mathbf{t})$, the output of this network is $\mathbf{o} = = (o_1, o_2, ..., o_c)$ where the i-th output value is

$$o_i = f(\beta_i - \theta_i). \tag{2.3}$$

Fig. 2.4 3-layer neural network with *a-b-c* configuration

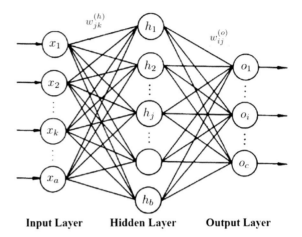

Thus, $(\mathbf{x}, \mathbf{t})$'s instantaneous error energy E that is produced at the output layer is

$$E = \frac{1}{2} \sum_{i=1}^{c} (t_i - o_i)^2. \tag{2.4}$$

Normally, there are $(d + l + 1)q + l$ unknown parameters need to be determined in this network. The standard BP algorithm is an iteration algorithm lasts several learning rounds. In each round, only a training instance $(\mathbf{x}, \mathbf{t})$ is processed for updating parameters. That is, each update value δ is generated according to the error $E(\cdot)$ in the current round. The BP algorithm is based on the *gradient descent* method, in which the adjustment of a parameter is proportional to the negative of the gradient of $E(\cdot)$ at the current instance $(\mathbf{x}, \mathbf{t})$. Given a learning rate η, the update values of weights are computed as follows.

$$w_{ij}^{(o)} \leftarrow w_{ij}^{(o)} - \eta \frac{\partial E}{\partial w_{ij}^{(o)}}. \tag{2.5}$$

$$w_{jk}^{(h)} \leftarrow w_{jk}^{(h)} - \eta \frac{\partial E}{\partial w_{jk}^{(h)}}. \tag{2.6}$$

We can derive that the partial derivatives are

$$\frac{\partial E}{\partial w_{ij}^{(o)}} = -(t_i - o_i)h_j; \tag{2.7}$$

$$\frac{\partial E}{\partial w_{jk}^{(h)}} = -h_j(1 - h_j)x_k \sum_{i=1}^{c} [(t_i - o_i)w_{ij}^{(o)}]. \tag{2.8}$$

Algorithm 1 Back-propagation Algorithm.

Input:
   N input vectors $\mathbf{x}$ with a variables and desired response vectors $\mathbf{t}$ with c variables, max iteration times max, learning rate η, and sigmoid function $f(\cdot)$

Output:
   neural network model with final weights: $w_{jk}^{(h)}$, $w_{ij}^{(o)}$, where $1 \le k \le a$, $1 \le j \le b$, and $1 \le i \le c$

1: Randomly initialize each weight $w_{jk}^{(h)}$, $w_{ij}^{(o)}$;

2: **for** $r = 1, 2, ..., max$ **do**
3:    **for** each instance vector $\mathbf{x}$ **do**
4:       // *Feed-forward Stage*
5:       **for** $j = 1, 2, ..., b$ **do**
6:          $h_j \leftarrow f(\sum_{k=1}^{a} x_k * w_{jk}^{(h)})$;
7:       **end for**
8:       **for** $i = 1, 2, ..., c$ **do**
9:          $o_i \leftarrow \sum_{j=1}^{b} h_j * w_{ij}^{(o)}$;
10:      **end for**
11:      // *Back-propagation Stage*
12:      Compute each $\frac{\partial E}{\partial w_{ij}^{(h)}}$ using (Eq. 2.7);
13:      Compute each $\frac{\partial E}{\partial w_{jk}^{(o)}}$ using (Eq. 2.8);
14:      Each $w_{ij}^{(o)} \leftarrow w_{ij}^{(o)} - \eta \frac{\partial E}{\partial w_{ij}^{(o)}}$;
15:      Each $w_{jk}^{(h)} \leftarrow w_{jk}^{(h)} - \eta \frac{\partial E}{\partial w_{jk}^{(h)}}$;
16:    **end for**
17: **end for**
18: **return** each $w_{jk}^{(h)}$, $w_{ij}^{(o)}$.

As shown in Algorithm 1, a learning round of the BP algorithm proceeds in two stages. In the *feed-forward* phase, the weights of the network are fixed and the input vector $\mathbf{x}$ is propagated through the network, layer by layer, until it reaches the output $\mathbf{o}$. Thus, changes are confined to the activation potentials and outputs in the network. In the *back-propagation* phase, an error value is produced by comparing the output of the network with a desired response $\mathbf{t}$. The resulting error value is propagated through the network, again layer by layer, but this time the propagation is performed in the backward direction. As discussed above, calculation of the adjustments for each layer is to get the corresponding partial derivatives. One full iteration over all available training data is an *epoch*. A learning epoch contains several learning rounds and is continued until the entire set of training instances has been accounted for. The training will last several epochs until the chosen stopping criterion (e.g., iteration $r = max$ in Algorithm 1) is met.

Normally, multiply trainers can joint hold a training dataset in different settings. We use a set of training instances $D = \{(\mathbf{x}^{(1)}, \mathbf{t}^{(1)}), ..., (\mathbf{x}^{(m)}, \mathbf{t}^{(m)})\}$ to describe two typical setting as follows.

- *Horizontally Partitioned Dataset.* In this setting, the sth trainer holds a subset D_s of D, where each training instance is in the form as $(\mathbf{x}, \mathbf{t})$. There are $D_1 \cup \cdots \cup D_N = D$ and $D_1 \cap \cdots \cap D_n = \emptyset$.
- *Vertically Partitioned Dataset.* In this setting, each trainer holds some components of a vector $\mathbf{x}$, and all trainers jointly hold all the set. For example, in a two-party

learning, for any $\mathbf{x} = (x_1, ..., x_a)$ where $a = a_1 + a_2$, trainer $\mathcal{A}$ holds $(x_1, ..., x_{a_1})$ and trainer $\mathcal{B}$ holds $(x_{a_1+1}, ..., x_{a_1+a_2})$. Each $\mathbf{t}$ is the desired output vector that is known to all the trainers.

Considering only one training instance is processed in a round of the BP algorithm, we easily give a privacy-preserving solution for the former setting. That is, a trainer builds the neural network model first and passes the training result to another trainer so that she can further train the model with her data, and so on. The final result is shared among all the trainers. So, the vertical-partition case is much more technically challenging. Generally, the challenge for designing a BP-based privacy-preserving learning scheme is how to finish two non-linear operations in a secure way. One is computing the *sigmoid function* in the feed-forward stage, and the other is computing *partial derivatives* in the back-propagation stage.

2.3 Vertically Partitioned Training Dataset

In the vertical-partition setting, Chen et al. (2009) proposed a privacy-preserving scheme based on the BP algorithm for the two-party learning. The main idea of this work is to protect the data confidentiality in each stage in Algorithm 1. It should ensure that neither the input data from the other trainer nor the intermediate results can be revealed.

2.3.1 Privacy-Preserving Two-Party Training

In this setting, two parties $\mathcal{A}$ and $\mathcal{B}$ jointly holds a set of training samples that are vertically partitioned. $\mathcal{A}$ owns one partition D_1 with $m_{\mathcal{A}}$ attributes for each data record. $\mathcal{B}$ owns the other partition D_2, with $m_{\mathcal{B}}$ attributes for each data record.

In each learning round, this privacy-preserving two-party neural network learning scheme enable two parties to jointly compute the additive values of connection weights without compromising their data privacy. Let training samples be $\mathbf{x}_{\mathcal{A}}$ and $\mathbf{x}_{\mathcal{A}}$ from $\mathcal{A}$ and $\mathcal{B}$ respectively and the corresponding label be $t_{\mathbf{x}}$. In more details, the goal of the scheme is to allow each party to get her own share of the additive value of each weight without revealing the original training data to each other.

Since inside adversaries are more powerful than outside adversaries (i.e., they have more background knowledge), the privacy concerns are considered to be brought by inside adversaries only. In practice, appropriate access control and security communication techniques can be used to defend outsider attacks such that all sensitive information is transmitted over secure channels and unauthorized access to the system is prevented.

2.3.2 Secure Manner

In the algorithms of privacy-preserving neural network learning, the ElGamal scheme is utilized to refresh ciphertexts. ElGamal is a public-key encryption scheme which can be defined on any cyclic group. Let G be a cyclic group of prime order q with generator g. Assume that decisional Diffie-Hellman assumption (DDH) holds in G such that ElGamal is semantically secure. ElGamal scheme consists of three components: key generation, encryption and decryption.

Key Generation. A value $x \in \mathbb{Z}_p$ is randomly chosen as the private key. The corresponding public key is $(G; q; g; h)$, where $h = g^x$.

Encryption. When a message $m \in G$ is encrypted, a value $r \in \mathbb{Z}_p$ is chosen as random. Then the ciphertext is constructed as $(C_1; C_2) = (g^r; m \cdot h^r)$.

Decryption. The plain text is computed as
$$\frac{C_2}{C_1^x} = \frac{m \cdot h^r}{g^{x \cdot r}} = \frac{m \cdot h^r}{h^r} = m.$$

Homomorphic Property. ElGamal scheme is multiplicative homomorphic for two messages. That is, an encryption of the multiplication of m_1 and m_2 can be obtained by an operation on $E(m_1; r)$ and $E(m_2; r)$ without decrypting any encrypted messages.

Probabilistic Property. ElGamal scheme is also probabilistic, whose probability is sourced from the a random value used for encrypting a clear-text message. For example, if an encryption of message m is encrypted by using the public key $(G; q; g; h)$ and a random value r, it is denoted as $E_{(G;q;g;h)}(m; r)$ ($E(m; r)$ for short). In probabilistic encryption schemes, there are many encryptions for each certain message m. ElGamal allows an operation that changes the random value used for encrypting a message without decrypting the corresponding ciphertext. This is called re-randomization or refresh operation. For instance, given an encrypted message $(C_1; C_2) = (g^r; m \cdot h^r)$, the re-randomization for obtaining another ciphertext of m is $(C_1'; C_2') = (C_1 \cdot g^s; C_2 \cdot h^s) = (g^{r+s}; m \cdot h^{r+s})$.

If messages are randomized by both parties, no individual party can decrypt any encryption by itself even though each party can partially decrypt an encryption. Alternatively, only if an encryption is partially decrypted by both parties, it is fully decrypted.

2.3.3 Scheme Details

To hide an intermediate result such as hidden layer values, the two trainers "randomly share" the result so that neither of them can imply the original training instances from it. This sharing method is each trainer holds a random number and the sum of the two random numbers equals to the intermediate result. Although intermediate results are randomly shared between the two trainers, the learning process can still securely carry on to produce the correct final result.

In a learning round, the current vector $\mathbf{x} = (x_1, ..., x_a)$ is vertically split into $\mathbf{x}_1 = (x_1, ..., x_{a_1})$ and $\mathbf{x}_2 = (x_{a_1+1}, ..., x_{a_1+a_2})$, where $a = a_1 + a_2$. Assuming that trainer $\mathcal{A}$ holds $\mathbf{x}_1$ and trainer $\mathcal{B}$ holds $\mathbf{x}_2$, the process for training current instance $(\mathbf{x}, \mathbf{t})$ is summarized in Algorithm 2.

This algorithm contains some secure computations for the "randomly share". These computations are based on two cryptographic tools. The one is ElGamal scheme (ElGamal 1985) with a property of cipher refresh, and the other is the partially decryption (Desmedt 1992). The cipher refresh means that ElGamal allows an operation that takes a message encrypted with a random factor as input and outputs a message encrypted with another random factor, but the two encrypted messages are encrypted from the same clear message. The key pairs of the encryption schemes are issued to $\mathcal{A}$ and $\mathcal{B}$ at the beginning of the training.

Secure Computation 1. It is to compute the sigmoid function between the two trainers. In order to apply cryptographic tools to non-linear functions like the sigmoid, this work chose a piecewise linear approximation of the activation function (Eq. 2.2) as follows.

$$
f(v) = \begin{cases}
1 & r - j, x > 8 \\
0.015625v + 0.875 & 4 < x \le 8, \\
0.03125v + 0.8125 & 2 < x \le 4, \\
0.125v + 0.625 & 1 < x \le 2, \\
0.25v + 0.5 & -1 < x \le 1, \\
0.125v + 0.375 & -2 < x \le -1, \\
0.03125v + 0.1875 & -4 < x \le -2, \\
0.015625v + 0.125 & -8 < x \le -4, \\
0 & x \le -8.
\end{cases}
\tag{2.9}
$$

The inputs can be seen as two integers which are x_1 and x_2 held by trainer $\mathcal{A}$ and trainer $\mathcal{B}$ respectively. The output $f(x_1 + x_2)$ is also randomly shared by $\mathcal{A}$ and $\mathcal{B}$. Although the trainers can learn the complete value of sigmoid function by exchanging their random shares, the shares can still keep secret since they are one-time intermediate results. The steps of this computation is shown as follows.

- Trainer $\mathcal{A}$ generates a random number R and computes $m_i = f(x_1 + i) - R$ for each $i \in \{-(n-1), ..., n\}$. $\mathcal{A}$ encrypts each m_i using the encryption $\mathcal{E}(\cdot)$ of ElGamal scheme and gets $\mathcal{E}(m_i; r_i)$, where each r_i is a new random number for the encryption. $\mathcal{A}$ sends each $\mathcal{E}(m_i; r_i)$ in the increasing order of i.
- Trainer $\mathcal{B}$ picks $\mathcal{E}(m_{x_2}; r_{x_2})$. She refresh it and sends $\mathcal{E}(m_{x_2}; r')$ back to $\mathcal{A}$, where $r' = r_{x_2} + s$ and s is only known to $\mathcal{B}$.
- Trainer $\mathcal{A}$ partially decrypts $\mathcal{E}(m_{x_2}; r')$ and sends the partially decrypted message to $\mathcal{B}$.
- Party $\mathcal{B}$ finally doing the other partial decryption of the partially decrypted message get $m_{x_2} = f(x_1 + x_2) - R$.

Algorithm 2 A Learning Round Two-party Privacy-preserving BP Algorithm.

Input:

current instance $\mathbf{x}_1$, $\mathbf{x}_2$, and $\mathbf{t}$;current weights $\{\{w_{jk}^{(h)}\}_{k=1}^{a}\}_{j=1}^{b}, \{\{w_{ij}^{(o)}\}_{j=1}^{b}\}_{i=1}^{c}$; learning rate η

Output:

modified weights $\{\{w_{jk}^{(h)}\}_{k=1}^{a}\}_{j=1}^{b}, \{\{w_{ij}^{(o)}\}_{j=1}^{b}\}_{i=1}^{c}$

1: // *Feed-forward Stage*
2: **for** $j = 1, 2, ..., b$ **do**
3:     Trainer $\mathcal{A}$ computes $\sum_{k=1}^{a_1} w_{jk}^{(h)} x_k$;
4:     Trainer $\mathcal{B}$ computes $\sum_{k=a_1}^{a_1+a_2} w_{jk}^{(h)} x_k$;
5:     Using *Secure Computation 1*, $\mathcal{A}$ and $\mathcal{B}$ jointly compute the sigmoid function and randomly share the output h_j.
        $\mathcal{A}$ receives h_{j1} and $\mathcal{B}$ receives h_{j2}.
6:     // There is $h_{j1} + h_{j2} = h_j = f(\sum_{k=1}^{a_1} w_{jk}^{(h)} x_k + \sum_{k=a_1}^{a_1+a_2} w_{jk}^{(h)} x_k)$;
7: **end for**
8: **for** $i = 1, 2, ..., c$ **do**
9:     $\mathcal{A}$ computes $o_{i1} \leftarrow \sum_{j=1}^{b} h_{j1} * w_{ij}^{(o)}$;
10:    $\mathcal{B}$ computes $o_{i2} \leftarrow \sum_{j=1}^{b} h_{j2} * w_{ij}^{(o)}$;
11:    // There is $o_{i1} + o_{i2} = o_i = \sum_{j=1}^{b} h_{j1} * w_{ij}^{(o)} + \sum_{j=1}^{b} h_{j2} * w_{ij}^{(o)}$;
12: **end for**
13: // *Back-propagation Stage*
14: **for** $i = 1, 2, ..., c$ **do**
15:    **for** $j = 1, 2, ..., b$ **do**
16:        Using *Secure Computation 2*, $\mathcal{A}$ and $\mathcal{B}$ jointly compute the product $h_{j1}o_{i2}$ and randomly share r_{11} and r_{12}
           respectively.
17:        $\mathcal{A}$ and $\mathcal{B}$ jointly compute the product $h_{j2}o_{i1}$ and randomly share r_{21} and r_{22} respectively.
18:        // There are $r_{11} + r_{12} = h_{j1}o_{i2}$ and $r_{21} + r_{22} = h_{j2}o_{i1}$;
19:        $\mathcal{A}$ computes $\delta_1 w_{ij}^{(o)} \leftarrow (o_{i1} - t_i)h_{j1} + r_{11} + r_{12}$;
20:        $\mathcal{B}$ computes $\delta_2 w_{ij}^{(o)} \leftarrow (o_{i2} - t_i)h_{j2} + r_{21} + r_{22}$;
21:    **end for**
22: **end for**
23: **for** $j = 1, 2, ..., b$ **do**
24:    **for** $k = 1, 2, ..., a_1$ **do**
25:        Using *Secure Computation 3*, $\mathcal{A}$ and $\mathcal{B}$ jointly share q_1 and q_2 respectively.
26:        Using *Secure Computation 2*, $\mathcal{A}$ and $\mathcal{B}$ jointly compute the product $x_k q_2$ and randomly share r_{61} and r_{62}
           respectively.
27:        $\mathcal{A}$ computes $\delta_1 w_{jk}^{(h)} \leftarrow q_1 x_k + r_{61}$;
28:        $\mathcal{B}$ computes $\delta_2 w_{jk}^{(h)} \leftarrow r_{62}$;
29:    **end for**
30:    **for** $k = a_1 + 1, ..., a_1 + a_2$ **do**
31:        Using *Secure Computation 3*, $\mathcal{A}$ and $\mathcal{B}$ jointly share q_1 and q_2 respectively.
32:        Using *Secure Computation 2*, $\mathcal{A}$ and $\mathcal{B}$ jointly compute the product $x_k q_1$ and randomly share r_{61} and r_{62}
           respectively.
33:        $\mathcal{A}$ computes $\delta_1 w_{jk}^{(h)} \leftarrow r_{61}$;
34:        $\mathcal{B}$ computes $\delta_2 w_{jk}^{(h)} \leftarrow q_2 x_k + r_{62}$;
35:    **end for**
36: **end for**
37: $\mathcal{A}$ (B, resp.) sends $\delta_1 w_{ij}^{(o)}$ and $\delta_1 w_{jk}^{(h)}$ ($\delta_2 w_{ij}^{(o)}$ and $\delta_2 w_{jk}^{(h)}$, resp.) to ($\mathcal{A}$, resp.);
38: $\mathcal{A}$ and $\mathcal{B}$ compute each $w_{ij}^{(o)} \leftarrow w_{ij}^{(o)} - \eta(\delta_1 w_{ij}^{(o)} + \delta_2 w_{ij}^{(o)})$;
39: $\mathcal{A}$ and $\mathcal{B}$ compute each $w_{jk}^{(h)} \leftarrow w_{jk}^{(h)} - \eta(\delta_1 w_{jk}^{(h)} + \delta_2 w_{jk}^{(k)})$;
40: **return** $\{\{w_{jk}^{(h)}\}_{k=1}^{a}\}_{j=1}^{b}, \{\{w_{ij}^{(o)}\}_{j=1}^{b}\}_{i=1}^{c}$.

It is clear that $m_{x_2} + R = f(x_1 + x_2)$ where R is only known to $\mathcal{A}$ and m_{x_2} is only known to $\mathcal{B}$.

Secure Computation 2. It is used for achieving the privacy-preserving production in Eqs. 2.7 and 2.8. The inputs can be seen as two integers which are M and N held by trainer $\mathcal{A}$ and trainer $\mathcal{B}$ respectively. The output $M \cdot N$ is the production randomly shared by $\mathcal{A}$ and $\mathcal{B}$. The steps of this computation is shown as follows.

- Trainer $\mathcal{A}$ generates a random number R and computes computes $m_i = M \cdot i - R$ for each $i \in \{-(n-1), ..., n\}$. $\mathcal{A}$ encrypts each m_i using the encryption $\mathcal{E}(\cdot)$ of ElGamal scheme and gets $\mathcal{E}(m_i; r_i)$, where each r_i is a new random number for the encryption. $\mathcal{A}$ sends each $\mathcal{E}(m_i; r_i)$ in the increasing order of i.
- Trainer $\mathcal{B}$ picks $\mathcal{E}(m_N; r_N)$. She refresh it and sends $\mathcal{E}(m_N; r')$ back to $\mathcal{A}$, where $r' = r_N + s$ and s is only known to $\mathcal{B}$.
- Trainer $\mathcal{A}$ partially decrypts $\mathcal{E}(m_N; r')$ and sends the partially decrypted message to $\mathcal{B}$.
- Party $\mathcal{B}$ finally doing the other partial decryption of the partially decrypted message get $m_N = M \cdot N - R$.

It is clear that $m_N + R = M \cdot N$ where R is only known to $\mathcal{A}$ and m_N is only known to $\mathcal{B}$.

Secure Computation 3. It is based on *Secure Computation 2* and used for privately computing partial derivatives of the hidden layers. The inputs are h_{j1}, h_{j2}, o_{i1}, and o_{i1}, which have been received by trainer $\mathcal{A}$ and $\mathcal{B}$ in Algorithm 2. The output is also randomly shared by $\mathcal{A}$ and $\mathcal{B}$. The steps of this computation is shown as follows.

- Using *Secure Computation 2*, trainer $\mathcal{A}$ and $\mathcal{B}$ jointly compute the product $h_{j1}h_{j2}$ and randomly share r_{31} and r_{32} respectively, where $r_{11} + r_{12} = h_{j1}o_{i2}$.
- Trainer $\mathcal{A}$ computes $p_1 \leftarrow h_{j1} - h_{j1}^2 - 2r_{31}$ and $s_1 \leftarrow \sum_{i=1}^{c}(-t_i + o_{i1})w_{ij}^{(o)}$. Trainer $\mathcal{B}$ computes $p_2 \leftarrow h_{j2} - h_{j2}^2 - 2r_{32}$ and $s_2 \leftarrow \sum_{i=1}^{c} o_{i2}w_{ij}^{(o)}$.
- Using *Secure Computation 2*, trainer $\mathcal{A}$ and $\mathcal{B}$ jointly compute the product s_1p_2 and randomly share r_{41} and r_{42} respectively, where $r_{41} + r_{42} = s_1p_2$.
- Using *Secure Computation 2*, trainer $\mathcal{A}$ and $\mathcal{B}$ jointly compute the product s_2p_1 and randomly share r_{51} and r_{52} respectively, where $r_{51} + r_{52} = s_2p_1$.
- Trainer $\mathcal{A}$ gets $q_1 \leftarrow s_1p_1 + r_{41} + r_{51}$. Trainer $\mathcal{B}$ gets $q_2 \leftarrow s_2p_2 + r_{42} + r_{52}$.

2.3.4 Analysis of Security and Accuracy Loss

The security of the algorithms can be easily carried out in the semi-honest model. In such a threat model, the parties follow the protocol and may utilize their own knowledge during the protocol execution. Moreover, parties are not allowed to learn anything beyond their outputs from the information they get throughout the protocol. As shown in Chen and Zhong (2009), a standard way is to construct a simulator which can simulate the view of each party in the protocol when this party is only given the input and outputs a result.

Both parties agree on a termination condition of training, so that the simulators can maintain the training loop of the protocol. In each training round, most of the message transmissions occur in the calls of where $\mathcal{B}$ and $\mathcal{A}$ respectively receive $\delta_1 w_{ij}^{(o)}$, $\delta_2 w_{ij}^{(o)}$ and $\delta_1 w_{jk}^{(h)}$, $\delta_2 w_{ij}^{(h)}$. The variable h_{j2} can be simulated because of the fact that the algorithm is secure (as shown above). The variables r_{12} and r_{22} can also be simulated based on the security of the algorithm. Since o_{i2} and t_i are part of the input of party $\mathcal{B}$, $\delta_2 w_{ij}^{(h)}$ can be simulated for $\mathcal{B}$ by only looking at its own input. Meanwhile, since weights are output of the training algorithm and η is known to both parties as input, by the weight update rule, $w_{ij}^{(o)} \leftarrow w_{ij}^{(o)} - \eta(\delta_1 w_{ij}^{(o)} + \delta_2 w_{ij}^{(o)})$ can be simulated for party $\mathcal{B}$. The simulation of $\delta_1 w_{jk}^{(h)}$ for party $\mathcal{B}$ is likewise. Similarly, the simulator for party $\mathcal{A}$ is constructed to simulate $\delta_2 w_{ij}^{(o)}$ and $\delta_2 w_{ij}^{(h)}$.

Error in truncation. There are two places in the algorithm where the approximation is introduced for the goal of privacy. One is that the sigmoid function used in the neuron forwarding outputs an approximation by using a piecewise linear function. Since cryptographic operations are on discrete finite fields, the intermediate results should be represented as real numbers during computing on ciphertexts. Thus, other necessary approximation is introduced by the transformation between real numbers and fixed-point representations, which enables the cryptographic operations. The influence of these two sources of approximation on the accuracy loss will be empirically evaluated. Then, a brief theoretical analysis of the accuracy loss caused by the fixed-point representations is shown as follows.

Suppose that the system, in which neural network is implemented, uses μ bits for representation of real numbers. The input numbers are pre-processed into finite field that is suitable for cryptographic operations. The μ-bit numbers can be truncated by chopping off the lowest γ bits and thus the new lowest order bit is unchanged. The precision error ratio can be bounded by $\epsilon = 2^{\mu-\gamma}$.

Therefore, in the feed forward stage, the error ratio bound introduced by number conversion for cryptographic operations is ϵ; In output-layer δ, the error ratio bound for $\delta_1 w_{ij}^{(o)}$ and $\delta_2 w_{ij}^{(o)}$ is $(1 + \epsilon)^2 - 1$; In hidden-layer δ, the error ratio bound for $\delta_1 w_{jk}^{(h)}$ and $\delta_2 w_{jk}^{(h)}$ is $(1 + \epsilon)^4 - 1$; In weight update, the update of weights introduces no successive error.

2.4 Arbitrarily Partitioned Training Dataset

Comparing to horizontal and vertical partition, the arbitrary partition is a more general partition type for trainers' data.

Arbitrarily Partitioned Dataset. Assume that there are N trainers and the whole set is $D = \{(\mathbf{x}^{(1)}, \mathbf{t}^{(1)}), ..., (\mathbf{x}^{(m)}, \mathbf{t}^{(m)})\}$. Each lth instance vector $\mathbf{x}^{(l)}$ in D has a components $(x_1^{(l)}, ..., x_a^{(l)})$. $D_s^{(l)}$ a subset of components of $\mathbf{x}^{(l)}$, which is held by the sth trainer. For each $l \in [m]$, there are $D_1^{(l)} \cup \cdots \cup D_N^{(l)} = \mathbf{x}^{(l)}$ and $D_1^{(l)} \cap \cdots \cap D_N^{(l)} = \emptyset$. When $|D_s^{(l)}| = \cdots = |D_s^{(l)}|$ and the components owned by a trainer in each instance are at the same position, the arbitrary partition becomes vertical partition. Similarly, it is horizontal partition if each trainer completely holds some $\mathbf{x}^{(l)}$.

In this setting, Yuan et al. (2014) proposed a privacy-preserving BP-based neural network learning scheme for the multi-party learning. The data confidentiality is achieved via the BGN homomorphic encryption (Boneh et al. 2005) that supports one multiplication and unlimited number of addition operations, which is suited for computations in neural network learning algorithms.

2.4.1 BGN Homomorphic Encryption

Homomorphic Encryption. Homomorphic encryption enables the computations of plaintexts to be performed on the corresponding ciphertexts without revealing the underlying plaintexts. A public-key encryption scheme $\mathcal{AHE}$ that supports addition operations is additively homomorphic. Given two encrypted messages $\mathcal{AHE}.Enc(a)$ and $\mathcal{AHE}.Enc(b)$ that are encrypted using the same public key, there exists a public-key operation $\odot$ such that $\mathcal{AHE}.Enc(a) \odot \mathcal{AHE}.Enc(b)$ is the encryption of $a + b$. Set $\mathcal{AHE}.Enc(a)$ be a ciphertext of an underlying plaintext a and c be a constant value. The multiplication ciphertext $\mathcal{AHE}.Enc(ca)$ can be implement by $\mathcal{AHE}.Enc(a) \oplus \cdots \oplus \mathcal{AHE}.Enc(a) = (\mathcal{AHE}.Enc(a))^c$.

Since the BGN scheme $\mathcal{BGN}$ supports one multiplication, given two encrypted messages $\mathcal{BGN}.Enc(a)$ and $\mathcal{BGN}.Enc(b)$, the ciphertext of ab can be computed. It is a public-key "doubly homomorphic" encryption scheme, which simultaneously supports one multiplication and unlimited number of addition operations. Therefore, given ciphertexts $\mathcal{BGN}.Enc(a_1)$, $\mathcal{BGN}.Enc(a_2)$, ..., $\mathcal{BGN}.Enc(a_i)$ and $\mathcal{BGN}.Enc(b_1), \mathcal{BGN}.Enc(b_2), ..., \mathcal{BGN}.Enc(b_i)$, one can compute $\mathcal{BGN}.Enc(a_1 \cdot b_1 + \cdots a_i \cdot b_i)$ without knowing the plaintext, where the ciphertexts are encrypted by the system's public key.

The BGN scheme is designed for two parties. Moreover, due to message decryption involves solving discrete logarithm of the ciphertext using Pollard's lambda method, BGN scheme just works with small numbers. While it is easy to make BGN decrypt long messages (in which the message is treated as a bit string) using a mode of operation, it is non-trivial to let BGN efficiently decrypt large numbers (wherein the final message is interpreted by its value and unknown to the encryptor after homomorphic operations).

2.4.2 Overviews

The learning system is composed of three kinds of entities, including a trusted authority (TA), several participants (data owners), and the outsourcing cloud servers. TA is the entity which has the responsibility for generating and issuing encryption/decryption keys for all the other entities in the system. For efficiency, it will not undertake in any task other than key generation and issuing. Each participant s is denoted as $\mathcal{T}_s$, which owns a private dataset and tries to perform collaborative BP network learning with each other participant. In another word, they will collaboratively and parallelly conduct learning over the joint dataset that are arbitrarily partitioned. Thus, any private part cannot be disclosed during the whole learning process.

The goal is to enable multiple participants to jointly conduct the network learning task by using the BP algorithm without revealing their private data. The computational and communicational costs on each participant shall be practically efficient for each participant in the scalable learning system. To achieve this goal, the intuition of this work is to implement a privacy-preserving mechanism on the non-privacy-preserving BP learning algorithm and thus make each step "privacy-preserving". Different from the original learning algorithm, each participant agrees on encrypting her/his input data set and uploading the encrypted data to the cloud server. Such a cloud server is supposed to perform most of the cipher operations, i.e., additions and scalar products.

To support these operations over ciphertexts, the BGN encryption is adopted and tailored for data encryption. Considering the BGN algorithm just supports one step multiplication over ciphertext, this work sets that the intermediate products or scalar products shall be first securely decrypted and then encrypted to support consecutive multiplication operations. Alternatively, if any participant knows the actual intermediate values, he/she will derive training data directly.

To achieve this, it is necessary to give a secret sharing algorithm that allows each participant to only obtain a random share of the intermediate values. It is supposed that the participants can forwarding learning processes by using these random shares without revealing any intermediate value. Thus, the data privacy can be well protected. The overall algorithm is the core of the privacy-preserving scheme. The secure computation 4 for secure random sharing, while the secure computation 5 for securely computing the sigmoid function approximation. After the cooperative privacy-preserving learning, all the participants can jointly establish a neural network that represents the distribution of the whole dataset without exposing any private data to each other.

2.4.3 Scheme Details

Similar as the scheme in Sect. 2.3, the trick of this scheme is also "randomly share", but it is among multiple trainers. Moreover, a trusted authority and a cloud server are introduced in the scheme to finish some learning computations. In a learning round, the current vector $\mathbf{x} = (x_1^{(l)}, ..., x_a^{(l)})$ is vertically split into $D_1^{(l)}, ..., D_N^{(l)}$ among each sth trainer $\mathcal{T}_s$. The process for training current instance $(\mathbf{x}, \mathbf{t})$ (without loss of generality, the instance without the superscript is used to represent any one instance in D) is summarized in Algorithm 3.

At the beginning of the training, a BGN encryption scheme $\mathcal{BGN}$ is initialized by a Trusted Authority (TA), and each trainer is issued her own secret key. The master

---

Algorithm 3 A Learning Round Multi-party Privacy-preserving BP Algorithm.

Input:

partitioned current instance $D_1, ..., D_N$ held by each sth trainer $\mathcal{T}_s$ respectively; current desirable response $\mathbf{t}$; current weights $\{\{w_{jks}^{(h)}\}_{k=1}^a\}_{j=1}^b, \{\{w_{ijs}^{(o)}\}_{j=1}^b\}_{i=1}^c, ..., \{\{w_{jks}^{(h)}\}_{k=1}^a\}_{j=1}^b, \{\{w_{ijs}^{(o)}\}_{j=1}^b\}_{i=1}^c$ held by each each sth trainer

Output:

modified weights $\{\{w_{jks}^{(h)}\}_{k=1}^a\}_{j=1}^b, \{\{w_{ijs}^{(o)}\}_{j=1}^b\}_{i=1}^c, ..., \{\{w_{jks}^{(h)}\}_{k=1}^a\}_{j=1}^b, \{\{w_{ijs}^{(o)}\}_{j=1}^b\}_{i=1}^c$ held by each each sth trainer

1: // *Feed-forward Stage*

2: **for** $j = 1, 2, ..., b$ **do**

3:    Using $\mathcal{BGN}$ and *Secure Computation 4*, each $\mathcal{T}_s$ computes random share ϕ_s for $\sum_{k=1}^a (\sum_{s=1}^N x_{ks}) \cdot (\sum_{s=1}^N w_{jks}^{(h)})$;

4:    Using *Secure Computation 6*, each $\mathcal{T}_s$ computes the sigmoid function random share h_{js} for $h_j = f(\sum_{s=1}^N \phi_s)$;

5: **end for**

6: **for** $i = 1, 2, ..., c$ **do**

7:    Using $\mathcal{BGN}$, *Secure Computation 4*, and *Secure Computation 5*, each $\mathcal{T}_s$ computes random share o_{is} for $o_i = f(\sum_{j=1}^b (\sum_{s=1}^N h_{js}) \cdot (\sum_{s=1}^N w_{ijs}^{(o)}))$;

8: **end for**

9: // *Back-propagation Stage*

10: **for** $i = 1, 2, ..., c$ **do**

11:    **for** $j = 1, 2, ..., b$ **do**

12:       Using $\mathcal{BGN}$ and *Secure Computation 4*, each $\mathcal{T}_s$ computes random share $\delta w_{ijs}^{(o)}$ for $\delta w_{ij}^{(o)} = (-(t_i - \sum_{s=1}^N o_{is})) \cdot (\sum_{s=1}^N h_{js})$;

13:    **end for**

14: **end for**

15: **for** $j = 1, 2, ..., b$ **do**

16:    Using $\mathcal{BGN}$ and *Secure Computation 4*, each $\mathcal{T}_s$ computes random share μ_s for $\mu = \sum_{i=1}^c (\sum_{s=1}^N o_{is} - t_i) \cdot (\sum_{s=1}^N w_{ijs}^{(o)})$;

17:    Using $\mathcal{BGN}$ and *Secure Computation 4*, each $\mathcal{T}_s$ computes random share ϑ_s for $\vartheta = \sum_{s=1}^N h_{js} \cdot (1 - \sum_{s=1}^N h_{js})$;

18:    **for** $k = 1, 2, ..., a$ **do**

19:       Using $\mathcal{BGN}$ and *Secure Computation 4*, each $\mathcal{T}_s$ computes random share κ_s for $\kappa = \sum_{s=1}^N x_{ks} \cdot \sum_{s=1}^N \mu_s$;

20:       Using $\mathcal{BGN}$ and *Secure Computation 4*, each $\mathcal{T}_s$ computes random share $\delta w_{jks}^{(h)}$ for $\delta w_{jk}^{(h)} = \sum_{s=1}^N \kappa_s \cdot \sum_{s=1}^N \vartheta_s$;

21:    **end for**

22: **end for**

23: Each $\mathcal{T}_s$ computes $w_{ijs}^{(o)} \leftarrow w_{ijs}^{(o)} - \eta \delta w_{ijs}^{(o)}$;

24: Each $\mathcal{T}_s$ computes $w_{jks}^{(h)} \leftarrow w_{jks}^{(h)} - \eta \delta w_{jks}^{(h)}$;

25: **return** $\{\{w_{jk1}^{(h)}\}_{k=1}^a\}_{j=1}^b, \{\{w_{ij1}^{(o)}\}_{j=1}^b\}_{i=1}^c, ..., \{\{w_{jkN}^{(h)}\}_{k=1}^a\}_{j=1}^b, \{\{w_{ijN}^{(o)}\}_{j=1}^b\}_{i=1}^c$.

---

secret key is only known to the TA. $\mathcal{BGN}$ is used for secure scalar productions and additions. $\mathcal{E}(\cdot)$ is the encryption algorithm of $\mathcal{BGN}$.

The key algorithm of this work allows multiple participants to perform secure scalar product and homomorphic addition operations on ciphertexts using cloud computing. For learning at each epochs, each party encrypts her/his data with the system public key and submits the ciphertexts to the cloud. The cloud servers carry out the sum of ciphertexts based on their ciphertexts. The cloud also computes the scalar product of the vectors if the original messages are vectors. Due to the homomorphic property and the semantic security of the cryptosystem, the cloud does not need to decrypt nor learn about the original messages during the cipher operations. The intermediate result of the sum or scalar product is returned to all the participants in ciphertext.

Decryption of Large Numbers. The work Yuan and Yu (2014) gives the method for decryption of large numbers. Each participant needs to use Pollard's lambda method to decrypt BGN ciphertexts. Due to efficiency limitation of the Pollard's lambda method, the algorithm can only work well with relative small numbers. The decryption algorithm of BGN depends on the solution of a discrete log problem, which is based on Pollard's lambda method. Thus, decryption of larger numbers is considered to be impractical in terms of the time complexity. Suppose the magnitude as follows: the Pollard's lambda method is able to decrypt numbers of up to 30–40 bits within a reasonable time slot. Moreover, we can hardly guarantee that the final results are always small enough for the Pollard's lambda method to efficiently decrypt in practice. The reason is either some components in the vectors are too large, or the vectors are of high dimension. To this end, the numbers should be split into several relatively small numbers, if they are large. The cloud server then decrypt these smaller number so that the final result can be recovered. Moreover, the decryption process can be parallelized or distributed for efficiency. If the cloud server can efficiently decrypt the result whose underlying plaintext is less than d bits, the scalar product operation can be considered over input numbers of $3d$ bits.

Therefore, instead of directly calculating the production between two large numbers, the participants can first compute three parts separately and finally recover one part. Then, the participants who hold the data need to split A_i and B_i and encrypt A_{i0}, A_{i1}, A_{i2} and B_{i0}, B_{i1}, B_{i2}, which are d-bit numbers. As a result, the computational cost for the encryption on each participant increases by x times, where x is the number of the partitions of each split large number ($x = 3$ in the example above). That for the scalar product and the decryption increases x^2 times and x^2 times. respectively. If x is a fixed small number, the expansion of computation/communication scale is constant and does not involve too much overhead.

This algorithm also contains some secure computations for the "randomly share".

Secure Computation 4. As discussed previously, the actual intermediate results should also be protected and cannot be known to each participant or the cloud server. When running the BP algorithm collaboratively, the participants need to execute consecutive operations such as addition and multiplication. Considering the BGN algorithm only supports one step multiplication over ciphertext, the participants need to share ciphertexts and decrypt the partial intermediate results. To avoid revealing intermediate values by a straightforward decryption, a secure sharing algorithm enables each participant to get a random share of the intermediate result without knowing its actual value.

The secure computation 4 is used for securely share a product or sum. To securely share the result ϵ, the steps of this computation should be finished as follows.

- Each trainer $\mathcal{T}_s$ generates a random number L_s and encrypts it to $\mathcal{E}(L_s)$.
- Each $\mathcal{T}_s$ uploads the $\mathcal{E}(L_s)$ to the cloud server, and the server computes the ciphertext of $sumL = \sum_{s=1}^{N} L_s$ using the cipher addition of $\mathcal{BGN}$.
- All the trainer work together to decrypt the difference between ϵ and $sumL$ as $L = |\epsilon - L|$ and send it to $\mathcal{T}_1$.

- Finally, each T_s gets the share ϵ_s for ϵ. For $T_s(s > 2)$, $\epsilon_s = L_s$. For T_1, if $\epsilon < sumL$, $\epsilon_1 = L_1 - L$; otherwise, $\epsilon_1 = L_1 + L$.

It is clear that ϵ_s is only known to the trainer T_s.

Secure Computation 5. It is to compute the sigmoid function among multiply trainers. In order to apply cryptographic tools to non-linear functions like the sigmoid, this work chose the Maclaurin series to approximate the activation function (Eq. 2.2) as follows:

$$f(v) = \frac{1}{2} + \frac{v}{4} - \frac{v^3}{48} + \frac{v^5}{480}. \tag{2.10}$$

As described above, using *Secure Computation 4*, each trainer T_s can joint secretly share a value v as v_s. To share v^2, each T_s computes the ciphertext of $v \cdot v_s$ using $\mathcal{E}(v)$ and $\mathcal{E}(v_s)$ and uploads it to the cloud server. Then, using *Secure Computation 4*, each T_s can obtain a random share for v^2. The computations of v^3 and v^5 can be easily extended from the computation of v^2. Therefore, the "randomly share" of $f(v)$ can be achieved.

There is also an accuracy loss in this privacy-preserving scheme. Similar to other privacy-preserving works, the place that introduces accuracy loss is the approximation of the activation function. Here, the Maclaurin series expansion is utilized to approximate the function, whose accuracy can be adjusted by modifying the number of series terms according to the system requirement. As shown in Yuan and Yu (2014), compared to the non-privacy-preserving BP network learning algorithm, it introduces about 1.3–2% more error rate in 9 series terms setting in experiments.

References

Boneh D, Goh E-J, Nissim K (2005) Evaluating 2-dnf formulas on ciphertexts. In: Theory of cryptography conference. Springer, pp 325–341

Chen T, Zhong S (2009) Privacy-preserving backpropagation neural network learning. IEEE Trans Neural Netw 20(10):1554–1564

Desmedt Y (1992) Threshold cryptosystems. In: International workshop on the theory and application of cryptographic techniques. Springer, pp 1–14

ElGamal T (1985) A public key cryptosystem and a signature scheme based on discrete logarithms. IEEE Trans Inf Theory 31(4):469–472

Yuan J, Yu S (2014) Privacy preserving back-propagation neural network learning made practical with cloud computing. IEEE Trans Parallel Distrib Syst 25(1):212–221

Chapter 3
Outsourced Computation for Learning

3.1 Outsourced Computation

During machine learning tasks, the trainers could suffer some bottlenecks on resources of computation, communication, and storage. With the rapid development of Internet services, cloud computing provide a solution for large scale computations, which is the so-called outsourcing paradigm. Through the outsourcing computations, a resource-constrained device can delegates its computation tasks to cloud servers in a pay-per-use manner. In this way, it enjoys the unlimited computation resources if they have enough bandwidths. Moreover, if we see the machine learning task as a special computation, the trainer that wants to build a model can avoid large cost on deployment and maintenance of learning algorithms by the outsourcing paradigm.

Although the outsourcing paradigm provides tremendous benefits, it is still inevitable that some new security concerns and challenges arise in the outsourcing computation. Firstly, the cloud server is not totally trusted, such that it is curious about sensitive information in the computation tasks. However, the sensitive information should not be revealed by the cloud servers. Thus, the outsourcing computation should ensure that the cloud servers cannot learn anything about the secret inputs and the outputs of the computation. Similar to other privacy-preserving applications, the encryption can only provide the confidentiality guarantees rather than meaningful computations over the encrypted data. Secondly, the cloud server may be malicious such that return some invalid computational results. Of course, this failure may be caused by a software bug or a lazy behaviour that decreases the amount of the computations due to financial incentives. Therefore, the result of the outsourcing computation should be checkable, which means any failure has to be detected. Obviously, it must be far more efficient than performing the computation task locally whether protecting data privacy or checking computational results.

Then, we will introduce a secure outsourcing scheme for machine learning tasks.

J. Li et al., *Privacy-Preserving Machine Learning*, SpringerBriefs on Cyber Security Systems and Networks, https://doi.org/10.1007/978-981-16-9139-3_3

3.2 Multi-key Privacy-Preserving Deep Learning

In the cloud environment, Li et al. (2017) present a basic outsourcing scheme based on multi-key fully homomorphic encryption (MK-FHE) along with an advanced scheme based on a hybrid structure by combining the double decryption mechanism and fully homomorphic encryption (FHE).

3.2.1 Deep Learning

The model of deep learning can be viewed as a neural network with the multi-layer perception structure. Like other multi-layer perceptions, the data records or variables are presented and input in the *input layer*. There are more than one *hidden layers*, which extract increasingly abstract features from the input layer. The "hidden" means that the parameters for these layers are not directly given in the data. During the learning process, it must be determined which features are useful for explaining the relationships of the input data. Each hidden layer takes a vector of real-valued as input, makes a linear combination of these vector and corresponding weight, and then outputs on each neuron by applying a non-linear activation function to the total input value. This calculation is defined as

$$x_k = f(x_{k-1}W_k + b) \tag{3.1}$$

where f is a predetermined activation function and W_k is a real-valued parameter matrix in hidden layer k (called *weight matrix*). A weight matrix determines the contribution of the input to the output in each layer. The parameter b is called *bias* which is usually used to ensure that the input sum is greater than 0. Algorithm 1 describes the back-propagation learning process of a multi-layer deep network and Fig. 3.1 depicts a neural network with two hidden layers for the binary classification.

There are many fashions of activation functions. One of the activation functions is the *logistic sigmoid* $f(z) = \frac{1}{1+exp(-z)}$. Usually, the logistic sigmoid function is used to produce a value ϕ with the range $(0, 1)$, which follows the Bernoulli distribution. Another activation function is *softplus* function $\zeta(z) = log(1 + exp(z))$, which can be used to produce a value with the range $(0, \infty)$, which follows the normal distribution.

Gradient-Based Optimization. The a deep learning task is to solve a optimization problem. The objective of such a optimization $f(\mathbf{x})$ is supposed to be minimized or maximized on its parameters. Hence, this function f is called *objective function*. This function is also called *error function*, *loss function*, or *cost function*, when the optimization is a minimization.

Assuming that there is a function $y = f(\mathbf{x})$, where $\mathbf{x}$ is vector and y is a real number. Let $f'(\mathbf{x})$ or $\frac{\partial y}{\partial \mathbf{x}}$ denote the *derivative* of this function f, its Taylor series expansion is

---

Algorithm 1 Multi-layer Back-propagation network learning

Require: input sample x, target vector t, learning rate η, sigmoid function $f(x)$, and network depth l;

Ensure: the weight matrices of the model: $W^{(i)}, i \in \{1, 2, \ldots, l\}$

1: Feed Forward Stage:

2: $h^{(0)} = x$;

3: **for** $k = 1$ to l **do**

4:    $v^{(k)} = W^{(k)}h^{(k-1)}$;

5:    $h^{(k)} = f(v^{(k)})$;

6: **end for**

7: $y = h^{(l)}$

8: $E = E(t, y)$//compute the cost function

9: Back-propagation Stage:

10: $e \leftarrow \nabla_y E = \nabla_y E(t, y)$//compute the gradients of E with parameter W

11: **for** $k = l$ to 1 **do**

12:    $e \leftarrow \nabla_{v^{(k)}} E = ef'(v^{(k)})$//convert the gradient on the layer's output into the gradient pre-nonlinearity activation

13:    $\nabla_{W^{(k)}} E = eh^{(k-1)\tau}$//compute the gradients on weights

14:    $e \leftarrow \nabla_h^{(k-1)} E = W^{(k)\tau} e$//modify the next lower-level hidden layer's activations

15: **end for**

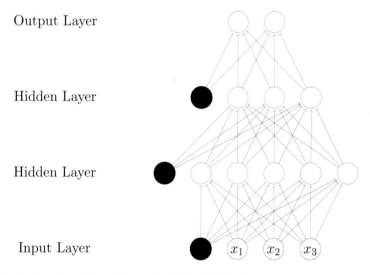

Fig. 3.1 Configuration of Neural Network with Two Hidden Layers

$$f(\mathbf{x} + \varepsilon) \approx f(\mathbf{x}) + \varepsilon \cdot f'(\mathbf{x}). \qquad (3.2)$$

The above Eq. (3.2) shows that scaling a small change ε in the input data $\mathbf{x}$ will cause a small change of y. Therefore, the derivative of parameters can be used for minimizing a function iteratively. This technique of changing parameters along the tangent orientation is called *gradient descent*. The condition $f'(\mathbf{x}) = 0$ means the tangent orientation at point $\mathbf{x}$ is 0, in which no information provided about what direction to move. Such points are called *critical points*, which can be either *local minimum* points or *global minimum* points of the function.

Stochastic gradient descent. The gradient descent algorithm is to solve two problems for achieving the optimization objective. They are the problem of the speed of the convergence to a local minimum and the problem in the error surface, since there are many local minima error does not mean that the global minimum error is found. Hence, to improve the convergence when updating parameters, an intuition for the optimization is to use a mini-batch of training samples at a time, i.e., the weights are not updated upon examining all training examples round and round. This optimization algorithm is called *stochastic* gradient descent (SGD), which can be seen as an extension of the gradient descent algorithm.

In general, the gradient in stochastic gradient descent is viewed as an estimation which approximately calculates the gradient of the whole training dataset by that of a subset of the dataset. In more details, assuming that n is the size of training dataset, a subset $X = \{X_1, \ldots, X_{n'}\}$ is chose as a mini-batch, which is uniformly sampled from the whole training dataset. The value n' is the size of the mini-batch, which is usually is fixed and far smaller than n. Algorithm 2 shows the detail of the SGD algorithm by taking the average gradient on a mini-batch of n training examples.

Algorithm 2 Stochastic gradient descent (SGD)

Require: Learning rate η
Ensure: Initial weight parameter w
  while stopping criterion not met **do**
    Sample a mini-batch of n' examples from the training sample set $\{X_1, \ldots, X_{n'}\}$ with the corresponding target outputs t_i.
    for $i = 1$ to n' **do**
      Calculate gradient estimate: $y \leftarrow \frac{1}{n'} \nabla_w \sum_i E(f(X_i; w); t_i)$;
    end for
    Apply update: $w \leftarrow w_k - \eta \cdot y$
  end while

As the setting in previous chapters, the learning rate η is set as a small positive scalar, which is used to control the step size during the gradient descent search. Here, the cost function (or the error function) is denoted as E, which represents essentially the difference between the target output of the network and the expected output of the network. The vector $\nabla_w E(\cdot)$ is comprised of components whose contents are the partial derivatives of E with respect to each of the w_j, where w_j is the component of weight vector w.

3.2.2 Homomorphic Encryption with Double Decryption Mechanism

The outsourcing scheme adopts a special public-key encryption scheme. In the public-key encryption scheme with a double decryption mechanism, there exists two independent decryption algorithms. The one is the user decryption algorithm, which takes the general private key as input. The other is the master entity decryption algorithm, which takes the master private key as input. The two decryption procedures ensure the decryption on any given ciphertext to be successful. The definition of the BCP scheme Bresson et al. (2003) is shown as follows.

Definition 3.1 (*BCP scheme*) The BCP scheme is a tuple of five algorithms including setup algorithm *Setup*, key-generation algorithm *KeyGen*, encryption algorithm *Enc*, user decryption algorithm *Dec* and master decryption algorithm *mDec*, which is defined as $\mathcal{E} = \{Setup, KeyGen, Enc, Dec, mDec\}$.

- $(pp, mk) \leftarrow Setup(1^\kappa)$: given a security parameter κ, let p, q, p', q' be distinct odd primes and $p = 2p' + 1$ and $q = 2q' + 1$. Set the bit-length as $|N| = |pq| = \kappa$. For a multiplication group $\mathbb{Z}_{N^2}^*$, the algorithm *Setup* chooses a random element $g \in \mathbb{Z}_{N^2}^*$ of order $pp'qq'$ such that $g^{p'q'} \equiv 1 + kN \pmod{N^2}$ for $k \in \{1, 2, \ldots, N - 1\}$. After this step, the algorithm's outputs are

  public parameter $pp = (N, k, g)$ and master private key $mk = (p', q')$.

- $(pk, sk) \leftarrow KeyGen(pp)$: choose a random element $a \in \mathbb{Z}_{N^2}$ and compute the user's public key $h = g^a \bmod N^2$, where the user's private key is $sk = a$. Finally the algorithm outputs (pk, sk).
- $(A, B) \leftarrow Enc_{(pp,pk)}(m)$: given a message $m \in \mathbb{Z}_N$ and pick a random element $r \in \mathbb{Z}_{N^2}$, and outputs the ciphertext (A, B), where $A = g^r \bmod N^2$ and $B = h^r(1 + mN) \bmod N^2$.
- $m \leftarrow Dec_{(pp,sk)}(A, B)$: given a ciphertext (A, B) and private key $sk = a$, it returns the message m as

$$m = \frac{\frac{B}{A^a} - 1 \bmod N^2}{N}.$$

or the special message "reject" if it is invalid ciphertext.
- $m \leftarrow mDec_{(pp,pk,mk)}(A, B)$: given a ciphertext (A, B), pubic key $pk = h$ and master private key mk, the user's private key $(sk = a)$ can be computed as

$$a \bmod N = \frac{h^{p'q'} - 1 \bmod N^2}{N} \cdot k^{-1} \bmod N.$$

In order to remove the random element $r \in \mathbb{Z}_{N^2}$, it's necessary to compute

$$r \bmod N = \frac{A^{p'q'} - 1 \bmod N^2}{N} \cdot k^{-1} \bmod N.$$

then compute $\tau = ar \bmod N$. Finally, it returns the massage m as

$$m = \frac{(\frac{B}{g^{\tau}})^{p'q'} - 1 \bmod N^2}{N} \cdot \zeta^{-1} \bmod N.$$

or the special message "reject" if it is invalid ciphertext, where k^{-1} and ζ^{-1} denote the inverse of $k \bmod N^2$ and $p'q' \bmod N^2$, respectively.

To simplify the notation, we use $Enc_{pk}(m)$ instead of $Enc_{pp,pk}(m)$ and use $Add(\cdot)$ to denote the addition-gate, i.e., $(\bar{A}, \bar{B}) := (A_1 \cdot A_2 \bmod N^2, B_1 \cdot B_2 \bmod N^2) = Add(C_1, C_2)$, where $\{C_i\}_{i=1}^2 = \{(A_i, B_i)\}_{i=1}^2$ is a ciphertext set.

This encryption scheme is a fully homomorphic encryption (FHE) allows one to compute the encrypted results of additions and multiplications of two plaintexts using the corresponding two ciphertexts directly without decrypting any ciphertext.

Generally speaking, a FHE system $\mathcal{E}^F$ is a tuple of four algorithms, which consists of key generation algorithm F.KeyGen, encryption algorithm F.Enc, decryption algorithm F.Dec, and evaluation algorithm F.Eval. For this evaluation algorithm F.Eval, given a circuit C, a public key pk_F, and any ciphertexts c_i is generated by $F.Enc_{pk_F}(m_i)$, outputs a refreshed ciphertext c^* such that $F.Dec_{sk_F}(c^*) = C(m_1, \ldots, m_n)$.

Suppose that circuit D_{Add_F} can handle addition gate, denoted by Add_F. If c_1 and c_2 are two ciphertexts that encrypt m_1 and m_2, respectively, under pk_F, then we can compute

$$c \leftarrow F.Eval(pk_F, D_{Add_F}, c_1, c_2)$$

which is a ciphertext under pk_F of $m_1 + m_2$. Similarly, circuit D_{Multi_F} can handle multiplication gate, denoted as $Multi_F$. For ciphertexts c_1 and c_2 as defined before,

$$c \leftarrow F.Eval(pk_F, D_{Multi_F}, c_1, c_2)$$

is a ciphertext under pk_F of $m_1 \times m_2$. Assume $[m] = F.Enc_{pk_F}(m)$, then $[x] +_F [y] = [x + y] = Add_F([x], [y])$ and $[x] \times_F [y] = [xy] = Multi_F([x], [y])$ denote the addition and multiplication computation of FHE, respectively.

The multi-key fully homomorphic encryption (MK-FHE) is defined as follows.

Definition 3.2 (*Multi-key fully homomorphic encryption*) For arbitrary circuit class C, a family of encryption schemes $\{\mathcal{E}^n = (MF.KeyGen, MF.Enc, MF.Dec, MF.Eval)\}_{n>0}$ is said to be a multi-key fully homomorphic encryption, if for every $n > 0$, $\mathcal{E}^n$ satisfies the following properties:

- $(pk_{MF}, sk_{MF}, ek_{MF}) \leftarrow MF.KeyGen(1^{\kappa})$: given a security parameter κ, outputs a public key pk_{MF}, a private key sk_{MF} and a public evaluation key ek_{MF}.

- $c \leftarrow \mathsf{MF.Enc}(pk_{\mathsf{MF}}, x)$: for a message x and public key pk_{MF}, this algorithm outputs a ciphertext c.
- $x' \leftarrow \mathsf{MF.Dec}(sk_{\mathsf{MF}1}, sk_{\mathsf{MF}2}, \ldots, sk_{\mathsf{MF}n}, c)$: given a ciphertext c and n private keys $sk_{\mathsf{MF}1}, \ldots, sk_{\mathsf{MF}n}$, this algorithm outputs a message x'.
- $c^* \leftarrow \mathsf{MF.Eval}((c_1, pk_{\mathsf{MF}1}, ek_{\mathsf{MF}1}), \ldots, (c_m, pk_{\mathsf{MF}m}, ek_{\mathsf{MF}m}), C)$: taken any boolean circuit $C \in \mathbf{C}$, any valid m key pairs $(pk_{\mathsf{MF}1}, ek_{\mathsf{MF}1}), \ldots, (pk_{\mathsf{MF}m}, ek_{\mathsf{MF}m})$, and any ciphertexts $c_1, \ldots, c_m$ as input, this algorithm outputs a refreshed ciphertext c^*.

The scheme should satisfied the following properties: the correctness of decryption and compactness of ciphertexts. That is to say, for any circuit $C \in \mathbf{C}$, the support of $\mathsf{MF.KeyGen}(1^\kappa)$: n key pairs $\{(pk'_{\mathsf{MF}t}, sk'_{\mathsf{MF}t}, ek'_{\mathsf{MF}t})\}_{t \in [1,n]}$ and its any subset of m key pairs $\{(pk_{\mathsf{MF}i}, sk_{\mathsf{MF}i}, ek_{\mathsf{MF}i})\}_{i \in [1,m]}$, and any valid ciphertext $c_i \leftarrow \mathsf{MF.Enc}(pk_{\mathsf{MF}i}, x_i)(i \in [1, m])$, the algorithm $\mathsf{MF.Eval}$ holds the properties:

Correctness of decryption. given a tuple of private key $sk_{\mathsf{MF}'1}, \ldots, sk_{\mathsf{MF}'n}$, a refreshed ciphertext c^*, the correct decryption is

$$\mathsf{MF.Dec}(sk_{\mathsf{MF}'1}, \ldots, sk_{\mathsf{MF}'n}, c^*) = C(x_1, \ldots, x_m),$$

where $c^* \leftarrow \mathsf{MF.Eval}((c_1, pk_{\mathsf{MF}1}, ek_{\mathsf{MF}1}), \ldots, (c_m, pk_{\mathsf{MF}m}, ek_{\mathsf{MF}m}), C)$.
Compactness of ciphertexts: let

$$c^* \leftarrow \mathsf{MF.Eval}((c_1, pk_{\mathsf{MF}1}, ek_{\mathsf{MF}1}), \ldots, (c_m, pk_{\mathsf{MF}m}, ek_{\mathsf{MF}m}), C)$$

be a refreshed ciphertext, the size of c^* is independent of the parameter m and the size of C, i.e., there is a polynomial f holds $|c^*| \leq f(\kappa, n)$, where $|c^*|$ is denoted as the size of c^*.

If $n = 1$, the Definition 3.2 is the standard definition of FHE scheme. Here, Add_{MF} is used to denote the secure addition gate, which is given the ciphertext c_1 and c_2 of plaintext m_1 and m_2 under public key $pk_{\mathsf{MF}1}$ and $pk_{\mathsf{MF}2}$ respectively, the server calculates the sum as $c_1 +_{\mathsf{MF}} c_2 = Add_{\mathsf{MF}}(c_1, c_2)$. Similarly, $Multi_{\mathsf{MF}}$ denotes the secure multiplication gate, the server calculates the products as $c_1 \times_{\mathsf{MF}} c_2 = Multi_{\mathsf{MF}}(c_1, c_2)$.

3.2.3 Basic Scheme

In a deep learning system, each data owner has a private dataset DB_i in which local resources are fully administrated by the data owner. In this multi-key system, there are n such data owners which are denoted by $P_1, \ldots, P_n$. Each data owner $P_i (i \in [1, n])$ has own a key pair (pk_i, sk_i), which contains a public key and private keys, where $i \in [1, n]$. Moreover, its local sensitive dataset DB_i has I_i attributes $\{X_1^{(i)}, X_2^{(i)}, \ldots, X_{I_i}^{(i)}\}$ and $DB_1 \cap DB_2 \cdots \cap DB_n = \Phi$. These data owners want to perform collaborative deep learning with each other. Due to the limitation of computational resources,

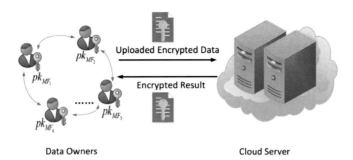

Fig. 3.2 The Basic Scheme

data owners need to outsource the data to an untrusted cloud server for finishing this collaborative learning task. Before running the learning algorithm, data owners agree on initiating an target vector $t = \{t_i\}_{i=1}^n$ and an weight matrix $W^{(j)}$ in advance, where $t_i = (t_1^{(i)}, t_2^{(i)}, \ldots, t_{l_i}^{(i)})$, $j = 1, 2$.

Because these datasets are highly sensitive, to preserve confidentiality of data, each data owner P_i ($i \in [1, n]$) has to encrypt his/hers data before uploading them to a cloud server. The cloud server will train a model over these encrypted datasets which are encrypted under *different public keys*. The privacy-preserving learning scheme ensures the inputs, intermediate results generated during the learning process and final output are secure and no information will be leaked during the whole learning process. In addition, each data owner is assumed to stay on-line with broadband access to the cloud server.

The system architecture of the basic scheme is depicted in Fig. 3.2, which realizes a scenario that multiple data owners want to collaboratively learn the parameters $W^{(1)}$, $W^{(2)}$ with their partitioned data without leaking the information of their sensitive datasets.

Intuition. Generally speaking, a straightforward privacy-preserving scheme cannot handle the data encrypted with different public keys. That is it can only deal with the ciphertext under the same public key, which is infeasible for an outsourcing computation task. Considering a multi-key scheme should preserve the data privacy when multiple parties are involved in deep learning model, a MK-FHE scheme $\mathcal{E}^n$ is utilized here to encrypt the data before uploading it to a cloud server. Assume there are n data owners $P_1, \ldots, P_n$ who hold their respect mutually disjoint, sensitive and vertically partitioned dataset $DB_1, \ldots, DB_n$, and corresponding target vectors $t_1, \ldots, t_n$. Each data owner P_i ($i \in [1, n]$) encrypts his dataset with MK.Enc and uploads the ciphertexts MK.Enc(pk_{MF_i}, DB_i), MK.Enc($pk_{MF_i}, W_i^{(j)}$), and MF.Enc(pk_{MF_i}, t_i) and its public key pk_{MF_i} to an untrusted cloud server $\mathcal{C}$ for the collaborative learning task.

To realize a multi-key privacy-preserving deep learning system, the concrete operation is shown in Algorithm 3. It is worth noting that the activation function $f(x) = \frac{1}{1+exp(-x)} \in (0, 1)$ in Algorithm 1 is nonlinear. To support effective compu-

---

Algorithm 3 Overall scheme of multi-key privacy-perserving deep learning based on MK-FHE

Require: $\{DB_1, DB_2, \ldots, DB_n\}$, initial $W^{(1)}$, $W^{(2)}$; $iteration_{max}$, Learning rate η
Ensure: $W^{(1)}$, $W^{(2)}$
1: Data owner $P_i (i \in [1, n])$ does:
2: Initialize the parameters randomly;
3: Sample a key tuple $(pk_{MF_i}, sk_{MF_i}, ek_{MF_i}) \leftarrow$ MF.KeyGen(1^κ);
4: Each data owner encrypts data DB_i, $W_i^{(j)}$: $c_i \leftarrow$ MF.Enc(pk_{MF_i}, DB_i), $d_i \leftarrow$ MK.Enc($pk_{MF_i}, W_i^{(j)}$), $g_i \leftarrow$ MK.Enc(pk_{MF_i}, t_i);
5: Upload $(pk_{MF_i}, ek_{MF_i}, c_i, d_i, g_i)$ to the cloud;
6: Cloud server dose:
7: Execute Algorithm 4 over the ciphertext domain;
8: Update the ciphertext of $W^{(j)}$;
9: Send the learning results τ^* to the data owners;
10: Data owners $P_1, \cdots, P_n$ do:
11: Data owners $P_1, \cdots, P_n$ jointly run a SMC protocol to calculate MF.Dec($sk_{MF_1}, \cdots, sk_{MF_n}, \tau^*$);

---

tation of the sigmoid function, an approximation of Taylor series is used here for the function, which is defined as:

$$\frac{1}{1 + e^{(-x)}} = \frac{1}{2} + \frac{x}{4} - \frac{x^3}{48} + o(x^4) \tag{3.3}$$

According to the property of Taylor series, the number of terms in the expansion can be decided according to the accuracy requirement. The secure computation of the activation function Eq. (3.3) is listed in Algorithm 4.

Algorithm 4 Securely outsourcing computation of activation function

---

Require: Ciphertexts $[x]$, $[a_0]$, $[a_1]$, $[a_3]$, where a_0, a_1, a_3 are constants.
Ensure: $[y]$
1: Compute c_0 with F.Enc(pk_F), i.e., $c_0 = [a_0]$;
2: Compute c_1 with secure multiplication $Multi_F$, i.e., $c_1 = [a_1] \times_F [x]$;
3: Compute c_3 with secure multiplication $Multi_F$, i.e., $c_3 = [a_3] \times_F [x] \times_F [x] \times_F [x]$;
4: Compute $[y]$ with secure addition Add_F, i.e., $[y] = c_0 +_F c_1 +_F c_3$;
5: **return** $[y]$;

3.2.4 Advance Scheme

Figure 3.3 illustrates an advance scheme that is more practical than the basic scheme. It finishes the multi-key privacy-preserving deep learning task without interaction among data owners, such that the communicational overhead is reduced. Three types

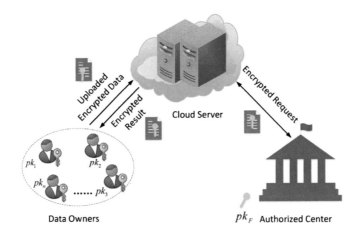

Fig. 3.3 The Advance Scheme

of entities are involved in this advanced scheme, including a cloud server C, an authorized center $\mathcal{AU}$, and n data owners $P_1, P_2, \ldots, P_n$.

Intuition. To decrease the computation/communication, Li et al. (2017) give a hybrid multi-key deep learning training system model. By using a double decryption mechanism and FHE, the training process can be performed over ciphertexts under different public key. Assume each data owner has uploaded the ciphertext (encrypted under its own public key of BCP scheme) to cloud server $\mathcal{C}$, which will perform a neural network with $(\alpha - \beta - \gamma)$ configuration over multi-key encrypted domain. For any valid ciphertext, authorized center $\mathcal{AU}$ (holds master key mk) can decrypt it by using the second decryption algorithm $mDec(\cdot)$ of the BCP scheme. Hence, $\mathcal{C}$ needs to blind the ciphertexts before sending them to the authorized center $\mathcal{AU}$. After receiving the blinded ciphertexts from the cloud $\mathcal{C}$, $\mathcal{AU}$ decrypts it by using the $mDec(\cdot)$ and re-encrypts the blinded plaintext by using FHE $\mathsf{F.Enc}_{pk_\mathsf{F}}(\cdot)$ of $\mathcal{E}^\mathsf{F}$. Finally, $\mathcal{AU}$ sends the fully homomorphic ciphertext to the cloud $\mathcal{C}$. As a result, the learning task can be performed over the re-encrypted data. Note that $\mathcal{C}$ and $\mathcal{AU}$ are honest-but-curios and do not collude with each other.

The details of the advanced scheme are described as follows.

Initialization. To protect the security and privacy of the data, each data owner encrypts his/her dataset before uploading to $\mathcal{C}$. The hybrid scheme is a combination of double decryption mechanism (BCP scheme) and FHE as our encryption technique. In the initialization process, $\mathcal{AU}$ sets up BCP scheme and FHE, uses $Setup$ of BCP scheme to generate a public parameter $pp = (N, k, g)$ and a master key $mk = (p', q')$. Then $\mathcal{AU}$ sends pp to $\mathcal{C}$ while keeping mk. The initialized parameters $W^{(1)}$, $W^{(2)}$ and target vector $\{T_i\}_{i=1}^n$ should be jointly negotiated in advance by the data owners. Note that $W^{(1)} = (W_1^{(1)}, \ldots, W_n^{(1)})^\tau$ and $W_i^{(1)} = (w_{ij}^{(1)})_{I_i \times \gamma}$. $W^{(2)} = (W_1^{(2)}, \ldots, W_n^{(2)})^\tau$, $\sum_{i=1}^n I_i = \alpha$; $W_i^{(2)} = (w_{ij}^{(2)})_{J_i \times \beta}$, $\sum_{i=1}^n J_i = \gamma$; $T = \{T_i\}_{i=1}^n$ is target vector and $T_i = (t_1, \ldots, t_{J_i})$, I_i and J_i is the number of input and output of P_i $(i \in [1, n])$, respectively.

Data uploading. In this phase, data owner $P_i (i \in [1, n])$ uses the received public parameter $pp = (N, k, g)$ to generate it own pair of public and private keys $\{(pk_i, sk_i)\}_{i=1}^n$. Assume that each data owner $P_i (i \in [1, n])$ has a set of vectors $DB_i = (X_1^{(i)}, X_2^{(i)} \ldots, X_{I_i}^{(i)})$. Data owner P_i $(i \in [1, n])$ uses BCP scheme to encrypt its private dataset $(Enc_{pk_i}(DB_i) = \{Enc_{pk_i}(X_j^{(i)})\}_{j=1}^{I_i} = \{(A_j^{(i)}, B_j^{(i)})\}_{j=1}^{I_i})$ and weight parameter $Enc_{pk_i}(W_i^{(1)})$, $Enc_{pk_i}(W_i^{(2)})$.

After that, data owner P_i $(i \in [1, n])$ uploads the encrypted data, the target vector T_i and the public key pk_i to the cloud server $\mathcal{C}$. Here, the communication channel for uploading is secure (can be implemented by some secure protocols), which means nobody can obtain the uploaded data from $\mathcal{C}$. For simplicity, (A_i, B_i) is used to denote BCP ciphertext of the DB_i for the data owner P_i $(i \in [1, n])$.

Training. The cloud server $\mathcal{C}$ should handles α request from data owners in each learning round. After receiving the data encrypted with *different public keys*, cloud $\mathcal{C}$ needs to perform an α-input deep learning algorithm over encrypted domain $\{(Enc_{pk_1}(DB_1), Enc_{pk_1}(W_1^{(k)}), T_1), \ldots, (Enc_{pk_n}(DB_n), Enc_{pk_n}(W_n^{(k)}), T_n)\}$, where $k = 1, 2$; $\alpha = \sum_{s=1}^n I_s$.

Considering the learning algorithm cannot process the ciphertext under the same public keys directly, the cloud server $\mathcal{C}$ first runs Algorithm 5 with authorized center $\mathcal{AU}$ to transform the ciphertexts under different public keys into ciphertext under the same public key. Since $\mathcal{AU}$ holds the master key mk, it can decrypt any given valid ciphertext by using the master decryption of BCP scheme. Hence, $\mathcal{C}$ needs to blind ciphertexts $\{(A_i, B_i)\}_{i=1}^n$ with a random message $\{r_i\}_{i=1}^n (r_i \in \mathbb{Z}_N^{I_i})$ before sending the ciphertexts to $\mathcal{AU}$. After receiving the blinded ciphertexts $\{(A_i', B_i')\}_{i=1}^n$, $\mathcal{AU}$ decrypts the blinded ciphertexts and re-encrypts the blinded plaintext z_i with FHE $\mathsf{F.Enc}_{pk_F}(z_i)$ of $\mathcal{E}^F$, which is denoted by Z_i, and sends this new ciphertext Z_i to $\mathcal{C}$. By removing the blinding factor r_i, $\mathcal{C}$ can get the ciphertext under the public key pk_F of $\mathsf{F.Enc}(\cdot)$ without knowing the underlying plaintext.

Algorithm 5 Transformation of BCP ciphertexts under $pk_1, pk_2, \ldots, pk_n$ to FHE ciphertexts under pk_F

Require: $((A_i, B_i), pk_i)$ for $i \in [1, n]$
Ensure: (A_i'', B_i'')
1: Cloud server $\mathcal{C}$ does:
2: choose randomness value $r_i \in \mathbb{Z}_N^{I_i}$, and compute the blinded ciphertexts $(A_i', B_i') \leftarrow$ $Add((A_i, B_i), Enc_{pk_i}(r_i))$;
3: send these blinded ciphertexts and pk_i to the authorized center $\mathcal{AU}$;
4: Authorized center $\mathcal{AU}$ does:
5: $z_i \leftarrow mDec_{pk_i, mk}(A_i', B_i')$ //$\mathcal{AU}$ holds the mk of BCP scheme;
6: $Z_i \leftarrow \mathsf{F.Enc}_{pk_F}(z_i)$//$\mathcal{AU}$ encrypts z_i with fully homomorphic encryption $\mathcal{E}^F$;
7: send Z_i to the authorized center $\mathcal{AU}$;
8: Cloud server $\mathcal{C}$ does:
9: $C_i' \leftarrow Add_F((Z_i), \mathsf{F.Enc}_{pk_F}(-r_i))$;

After performing the Algorithm 5, the cloud server $\mathcal{C}$ holds the data encrypted with the *same public key* pk_F of F.Enc. Later $\mathcal{C}$ can realize deep learning over the encrypted domain.

Here, DB denotes the set of $\{DB_i\}_{i=1}^n$. $net^{(2)}$ and $net^{(3)}$ represents the input and output values of the hidden layer, respectively. $o^{(2)}$ and $o^{(3)}$ represents the activation and output layer values, respectively.

The cloud server $\mathcal{C}$ needs to compute the values of $net^{(2)}$, $net^{(3)}$, $o^{(2)}$ and $o^{(3)}$ over the encrypted domain. In the domain of plaintexts, there is

$$o^{(2)} = f(net^{(2)}) = f(W^{(1)} \cdot DB),$$
$$o^{(3)} = f(net^{(3)}) = f(W^{(2)} \cdot o^{(2)}).$$

Algorithm 4 can be used to compute the activation function over the encrypted domain as follows:

$$[o^{(2)}] = f([net^{(2)}]) = f([W^{(1)}] \times_F [DB]),$$
$$[o^{(3)}] = f([net^{(3)}]) = f([W^{(2)}] \times_F [o^{(2)}]).$$

The stochastic gradient descent over the plaintext domain (Algorithm 2) is shown as follows:

$$\delta_k^{(3)} = o_k^{(3)}(1 - o_k^{(3)})(t_k - o_k^{(3)}), \tag{3.4}$$

$$\delta_h^{(2)} = o_k^{(2)}(1 - o_k^{(2)})(\sum_{k \in D} W_{kh}\delta_k^{(3)}). \tag{3.5}$$

where $\delta_k^{(3)}$ and $\delta_h^{(2)}$ are the error terms for each network output unit k and for each hidden unit h, respectively. $\mathcal{D}$ denotes all of units whose immediate inputs including the output of unit h.

Finally, update each network weight $W^{(1)}$, $W^{(2)}$ by computing

$$W^{(2)} := W^{(2)} + \eta\delta_k^{(3)}x_{kh} \tag{3.6}$$

$$W^{(1)} := W^{(1)} + \eta\delta_h^{(2)}x_{hi}. \tag{3.7}$$

From the Eqs. (3.4) and (3.5) over plaintext domain, the ciphertext can be securely computed by Add_F and $Multi_F$ as $[\delta_k^{(3)}] = [o_k^{(3)}] \times_F ([1] +_F [o_k^{(3)}]) \times_F ([t_k] +_F [o_k^{(3)}])$ and $[\delta_h^{(2)}] = [o_k^{(2)}] \times_F ([1] +_F [o_k^{(2)}]) \times_F (\sum_{k \in D}[W_{kh}] \times_F [\delta_k^{(3)}])$ respectively. Since the ciphertexts $[x_{hi}]$, $[x_{kh}]$ and $[\eta]$ are known, $[\delta_k^{(3)}]$ and $[\delta_h^{(2)}]$ are securely computed, the Eqs. (3.6) and (3.7) can be performed over encrypted domain.

Extraction. After running the deep learning model, the cloud server $\mathcal{C}$ obtains a set of learning results $[\tau]$. Then $\mathcal{C}$ runs Algorithm 6 with $\mathcal{AU}$ to transform the data encrypted with pk_F into ciphertexts under the data owners' public keys $pk_1, \ldots, pk_n$.

Algorithm 6 Transformation of the FHE ciphertext under pk_F to BCP ciphetext under different public keys

---

Require: A ciphertext C under pk_F and $pk_1, pk_2, \ldots, pk_n$
Ensure: $\{(A_i, B_i)\}_{i=1}^{n}$
1: Cloud server $\mathcal{C}$ does:
2: randomly choose $r \in \mathbb{Z}_N$, and compute $D \leftarrow Add_F(C, \text{F.Enc}_{pk_F}(r))$;
3: send blinded ciphertext D to the authorized center $\mathcal{AU}$.;
4: Authorized center $\mathcal{AU}$ does:
5: $z \leftarrow \text{F.Dec}_{sk_F}(D)$;
6: $(Z_i, D_i) \leftarrow Enc_{pk_i}(z)$ for all $i \in [1, n]$;//*Encrypting ciphertexts by using Enc of BCP scheme*
7: send (Z_i, D_i) to the cloud server $\mathcal{C}$;
8: Cloud server $\mathcal{C}$ does:
9: $(A_i, B_i) \leftarrow Add((Z_i, D_i), Enc_{pk_i}(-r))$;//*Removing the blinding factor*

Finally, $\mathcal{C}$ sends these ciphertexts to the data owners $P_1, \ldots, P_n$, each of which can decrypt this ciphertext with its private key sk_i.

3.2.5 Security Analysis

The system aims to achieve the privacy of input data, the intermediate results security and output results security of the deep learning under the *semi-honest* model, i.e., all participants in our two schemes are assumed to be honest-but-curious. The definition of the semantic security (SS), which is equivalent to the security against the polynomially indistinguishable chosen-plaintext attack (referred as IND-CPA security), is described as follows.

Definition 3.3 (*Semantic Security*) A public-key encryption scheme $\mathcal{E} = (KeyGen, Enc, Dec)$ is semantically secure, if for any stateful PPT adversary $A = (A_1, A_2)$, its advantage $\text{Adv}_{\mathcal{E},A}^{SS}(k) := |Pr[\text{Exp}_{\mathcal{E},A}^{SS}(k) = 1] - \frac{1}{2}|$ is negligible, where the experiment $\text{Exp}_{\mathcal{E},A}^{SS}(k)$ is defined as follows:
$\text{Exp}_{\mathcal{E},A}^{SS}(k)$:
  $b \leftarrow \{0, 1\}$
  $(pk, sk) \leftarrow KeyGen(1^k)$
  $(m_0, m_1) \leftarrow A_1(pk)$
  $c \leftarrow Enc_{pk}(m_b)$
  $b' \leftarrow A_2(c)$
  If $b' = b$, return 1;
  else, return 0

Here, it is required that the two plaintexts have the same length. If they are not, padding message can be applied.

Privacy of Data. To protect the data privacy of the data owners, in the basic scheme and the advanced scheme, the MK-FHE scheme and the hybrid encryption scheme are used to support the secure computation, respectively. According to the definition of semantic security, there are conclusions as follows.

Corollary 3.4 (MK-FHE semantic security) *If the underlying encryption scheme is semantically secure, then the multi-key fully homomorphic encryption is semantically secure.*

Proof (*Sketch*) Let's assume the public key encryption scheme $\mathcal{E} = \{KeyGen, Enc, Dec\}$ is semantically secure. Based on this scheme $\mathcal{E}$, the challenger constructs a evaluate algorithm $Eval$, such that the new public key encryption scheme $\mathcal{E}' = \{KeyGen, Enc, Dec, Eval\}$ keeps homomorphic of addition and multiplication operations. If the evaluation key ek is public, then the adversary can compute $Eval$ directly according to the public key pk, the ciphertext c and the evaluation key ek. Therefore, the MK-FHE scheme is semantically secure.   □

Recall that in our *basic scheme*, data owners do not communicate with each other until the decryption phase. Each data owner $P_i (i \in [1, n])$ generates its own key tuple $(pk_{\mathsf{MF}_i}, sk_{\mathsf{MF}_i}, ek_{\mathsf{MF}_i})$ and encrypts its input DB_i under the public key pk_{MF_i} of MK-FHE. A *semi-honest* data owner P may collude with some data owners, and wants to reveal a sample vector DB uploaded by other data owners. However, data owners do not need to coin-flip for each other's random coins. From the Definition 3.2, it guarantees our *basic scheme* is secure against corrupt data owners. Therefore, the privacy of the data owners is confidential.

In our *advanced scheme*, because the BCP scheme and FHE are semantically secure (the detailed proof Bresson et al. (2003), Theorem 11), the cloud server $\mathcal{C}$ should be probabilistic polynomially bounded, and sends the blinded ciphertexts to $\mathcal{AU}$ for computation. The computing power of $\mathcal{AU}$, on the other hand, dose not have to be bounded, since it only receives the blinded ciphertexts and only be able to see the blinded messages, which are decrypted by master decryption algorithm of BCP scheme. Hence, the cloud server $\mathcal{C}$ and authorized center $\mathcal{AU}$ cannot obtain the learning results. Therefore, the privacy of the learning results is confidential and the following lemma is obtained.

Lemma 3.5 *Without any collusion, Algorithms 5 and 6 is privacy-preserving for the weight $W^{(1)}, W^{(2)}$.*

Privacy of Training Model. An honest-but-curious cloud server $\mathcal{C}$ can train a deep learning model privately. Because Add_F and $Multi_F$ are both semantically secure, there is no information leakage for $\mathcal{C}$. Hence, for the weights $W^{(1)}, W^{(2)}$ in the training process of *In the feed work stage* and *In the back propagation stage*, cloud server $\mathcal{C}$ performs Add_F and $Multi_F$ operations, therefore privacy for the whole training process is guaranteed.

References

Bresson E, Catalano D, Pointcheval D (2003) A simple public-key cryptosystem with a double trapdoor decryption mechanism and its applications. In: International conference on advances in cryptology-asiacrypt, pp 37–54

Li P, Li J, Huang Z, Li T, Gao C-Z, Yiu S-M, Chen K (2017) Multi-key privacy-preserving deep learning in cloud computing. Future Gen Comput Syst 74:76–85

Chapter 4
Secure Distributed Learning

Like the common setting of cooperative learning, each participant securely maintains a private dataset on its local device. All participants want to jointly train a global machine learning model on the union of all their dataset, such that they can make full use of the advantage of each dataset. Maybe, they need a central server to coordinate the learning task in a high-efficient way.

The privacy concern arises as the scenario mentioned in previous chapters. The reason is that each participant (including the central server) is not totally trusted for each other but wants to protect sensitive information in its private dataset. Moreover, it is desirable that if each participant can finish local computations parallelly rather than in a synchronous way, the cooperative training can be implemented more efficient. In this chapter, we will introduce a distributed training scheme for privacy-preserving machine learning tasks. Such a learning technique is called *federated learning*. And we will further introduce a technique for improving federated learning.

4.1 Distributed Privacy-Preserving Deep Learning

Shokri and Shmatikov (2015) proposed a distributed learning scheme that enables multiple participants to jointly learn an accurate deep learning network model without revealing their training datasets. This system allows participants to independently converge to a set of parameters and avoid over-fitting the local model to a single participant's training dataset.

Figure 4.1 illustrates the architecture of this deep learning system. Assume that there are N participants each of which holds a local private dataset available for training. This learning system sets up a central parameter server whose responsibility is to maintain the deep learning network model with the latest parameters and make it available to all participants. The i-th participant $\mathcal{T}_i$ maintains a local neural network model with parameters represented by a vector $\mathbf{w}^{(i)}$, while the server maintains a

© The Author(s), under exclusive license to Springer Nature Singapore Pte Ltd. 2022   47
J. Li et al., *Privacy-Preserving Machine Learning*, SpringerBriefs on Cyber Security Systems and Networks, https://doi.org/10.1007/978-981-16-9139-3_4

Fig. 4.1 Distributed
Privacy-Preserving Deep
Learning System

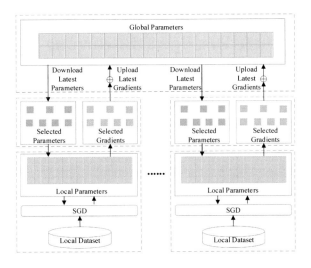

global parameter vector $\mathbf{w}^{(global)}$. Each participant initializes its local parameters and then executes the training on its own dataset. In each epoch, a participant locally performs the training based on Stochastic Gradient Descent (SGD). The coordination of this system is implemented by a *parameter exchange protocol* which allows participants to selectively submit their gradients of network parameters to the parameter server and selectively obtain the latest parameters at each epoch. Moreover, once the deep neural network model is trained, each trainer can independently perform secure evaluations of new data records.

4.1.1 Distributed Selective SGD

Considering some parameters contribute much more to the optimization of the objective function, it is intuitive that the global model can selectively updates parameters. The gradient values are carried out by the SGD algorithm works the training data in a mini-batch. Moreover, if a feature of the training data are more important than others, the parameters which is correlated to the feature will play a key role in the process of learning and make the convergence of the global model faster. Therefore, the selective SGD algorithm is used for training the global model. The "selective" here means the participant chooses a fraction of parameters and updates them using SGD at each iteration. This selection can be either completely random or specified by a heuristic strategy. Shokri and Shmatikov (2015) give a strategy that selects the parameters whose current values are farther away from their local optima. For each i-th sample in the j-th participant training dataset, the participant locally compute the partial derivative $\frac{\partial E_i}{\mathbf{w}_j}$ for all parameters $\mathbf{w}_j$ as in SGD. Let S be the indices of parameters with the $|S|$ largest $\frac{\partial E_i}{\mathbf{w}_j}$ values. In the end of each iteration, the participant updates the parameter vector $\mathbf{w}_S$ such that the parameters not in S remain unchanged.

The *distributed selective SGD* algorithm assumes multiple participants training independently and concurrently. Once a participant works out its local gradients for some of the parameters, it asynchronously share them with each other. Each participant can decide how often to share gradients and which to share. The magnitude of the global descent is determined by the sum of all gradients computed for a given parameter towards the parameter's local optima. Thus, participants can benefit from each other's training data and jointly produce more accurate models since they can avoid the overfitting on any local dataset. A central parameter server can help to exchange gradients among participants. It can hide the plaintext of each update by some security manners. By this way, participants asynchronously upload the gradients to the central server. The server aggregates all gradients to the value of the corresponding parameter. Each participant can download a subset of the parameters from the server and uses these parameters to update his local model further. In this work, the download criterion for a given parameter can be the moving average of gradients added to that parameter.

4.1.2 Scheme Details

Local Training. In summary, each participant locally trains a neural network model using the standard SGD algorithm iterating several epochs over her training data. Through the parameter server, participants influence each other's training indirectly. This procedure consists of five steps as follows in each learning epoch.

- The participant $\mathcal{T}_i$ downloads a θ_d fraction of parameters from the server and copies these downloaded values to the local model.
- $\mathcal{T}_i$ performs one epoch of SGD training over the local dataset.
- $\mathcal{T}_i$ computes $\delta\mathbf{w}^{(i)}$ which is the modification of the vector of all local parameters.
- $\mathcal{T}_i$ truncates each component of $\delta\mathbf{w}^{(i)}$ into the $[-\gamma, \gamma]$ range.
- $\mathcal{T}_i$ uploads $\delta\mathbf{w}_S^{(i)}$ to the parameter server, where S is the set of indices of at most $\theta_u \times |\mathbf{w}^{(i)}|$ gradients. As described previously, Such a set is selected according to one of the criteria: (i) a random subset of values that are larger than a threshold τ; (ii) exactly θ_u fraction of values that significantly contribute to the objective function.

Parameter Maintain. The parameter server maintains the parameter vector $\mathbf{w}^{(global)}$ of the global model and processes the each participant's upload and download requests as follows.

- *Upload* $\delta\mathbf{w}_S$. For each $j \in S$, the server adds the uploaded $\delta\mathbf{w}_j$ value to $\mathbf{w}^{(global)}$ and updates the counter $stat_j$.
- *Download* θ parameters. By setting θ_d, the trainer obtains the latest values of the parameters $\mathbf{w}_{I_\theta}^{(global)}$ with the largest $stat$ values I_θ.

The work shows that distributed parameter maintain in SGD can achieve almost the same accuracy as conventional and non-privacy-preserving SGD since it does not change the stochasticity of the learning process of SGD. Moreover, partially updating local parameters with global parameters during training will increase the stochasticity of local SGD. It is an essential way to prevent the local model from overfitting to each participant's local dataset, while trainer could fall into local optima when training alone. Local parameters are overwritten with the global parameters updated by other participants with different datasets helps each participant escape local optima and results in more accurate models.

Note that this distributed algorithm specifies neither the update rate nor which parameters need to be updated by other participants. Due to better computation and throughput capabilities, some trainers may undertake updates more frequently. Alternatively, some trainers may drop out or fail to upload their local parameters when suffering network errors or other failures. They may also overwrite each other's updated values since they are allowed to get access to the parameter server asynchronously. Instead of crippling the distributed SGD, such race conditions contribute to its success by increasing stochasticity. As a result, the asynchronous parameter update is effective for avoid overfitting and training accurate deep neural networks, as well as regularizing techniques that randomly corrupt neurons or input data during training.

Parameter Exchange Protocol. Participants do not need to follow any particular schedule when uploading their parameters. They can indirectly exchange parameters by one of the follow methods.

- Each participants downloads a fraction of the latest parameters from the parameter server, performs local training, and uploads selected gradients. Then, the next trainer follows in fixed order.
- With random order, participants download, training, and upload in random order. They have access to the server by locking it before downloading and releasing the lock after uploading.
- When a participant is training over a set of parameters, others may update these parameters on the server before training finishes.

Measure for Preserving Privacy. The measure is trivial but effective. Different from the traditional conventional deep learning, participants here do not reveal their training datasets to any others. That is, each participant learns the model and uses it locally and privately, without any interaction with others. Moreover, the selections of local datasets are kept confidential, and different training data can be used in each epoch. Therefore, the learning system can ensure the data privacy.

Also, for a stronger privacy guarantee, the differential privacy can be achieved in this learning system to ensure that parameter updates do not leak too much information about any individual in the whole dataset. We will talk about this and show the differentially privacy mechanism of this work in the chapter of *differential privacy*.

4.2   Secure Aggregation for Deep Learning

In a *federated learning* setting, some primitives in deep learning algorithms are relatively high-interaction. When training a neural network model, each trainer has to interact with others to finish these high-interaction operations and undertakes other works locally. One important primitive is computing a sum of data held by the trainers, which is referred to as the *aggregation*. For the server that maintains the aggregation among the trainers, it does not need to access any trainer's update values (i.e., may be the gradients) in order to perform SGD to update the global neural network model. That is, it requires only averages of the update values over a random subset of trainers.

In the federated learning setting, although each updated value is selected and contains no insignificant information about participants' local datasets, it may still result in the leakage of a local dataset. If the updated values are exposed to any one, it is possible to infer whether an individual record has been used for a participant's most recent update. Since the server only needs to perform aggregatively operations on the outputs of local SGD rather than get access to each individual participant's dataset for running SGD, parameters of the global model are weighted averages of the update vectors taken over a random subset of participants. Therefore, the secure aggregation can be implemented by the additively homomorphic encryption scheme. However, sometimes, the participant's device have only sporadic access to power and network connectivity, so the set of participants in each update step is unpredictable. It requires that the learning system must be robust to any participant's dropping out. However, a deep learning network may be very large (has millions of parameters), which makes participant maintaining very difficult. Moreover, a local device also generally cannot establish direct communications channels with other devices so as to build an authentication network for each other.

If neither the server nor the trainers can reveal the updated values in the clear but the aggregation still works, a privacy-preserving learning scheme can be thus constructed. Bonawitz et al. (2017) presented a scheme to securely aggregate data. The secure protocol in this scheme allows a server to compute the sum of data vectors held by participants without knowing any individual vector in the clear. Moreover, it can handle dropped participants. When training a neural network model, the server's task is to route messages between the other participants and compute the final result.

The system architecture of this scheme is shown in Fig. 4.2. The entities in this scheme is divided into two classes: a single server S that aggregates inputs from N participants $\mathcal{T}$. Each participant can be seen as a trainer $\mathcal{T}_u \in \mathcal{T}$ that holds a private feature vector $\mathbf{x}_u$ of dimension m. The components of $\mathbf{x}_u$ and $\sum_{u=1}^{N} \mathbf{x}_u$ are in $\mathbb{Z}_R$ for some R. The secure protocol of the scheme aims at computing $\sum_{u=1}^{N} \mathbf{x}_u$ in a secure fashion.

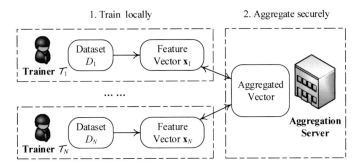

Fig. 4.2 System architecture of secure aggregation scheme

4.2.1 Secure Manner

The secure manner in the protocol is based on some cryptographic primitives, including Shamir's secret sharing, key agreement, authenticated encryption, and signature Scheme.

Secret Share. A t-out-of-n secret sharing scheme allows a secret s to be split into n shares. The scheme ensures that any t shares can be used to reconstruct s but any set of at most $t - 1$ shares gives no information about s.

The parameters of the scheme is over a finite field $\mathbb{F}$ (e.g., $\mathbb{F} = \mathbb{Z}_p$ for some large public prime p) of size at least $l > 2k$, where k is the security parameter of the scheme. The size of this field size needs to be large, because this scheme requires participants to secret share their secret keys whose length must depend on the security parameter for the security proof to go through. The participants' IDs $1, ..., n$ can be identified with distinct field elements in $\mathbb{F}$.

A secret share scheme $\mathcal{SS}$ consists of two algorithms ($\mathcal{SS}.Share, \mathcal{SS}.Recon$). The algorithm $\mathcal{KA}.Share$ $(s, t, \mathcal{T}) \rightarrow \{(\mathcal{T}_u, s_u)\}_{u \in [|\mathcal{T}|]}$ produces a set of shares s_u, each of which is associated with a different $\mathcal{T}_u \in \mathcal{T}$. $\mathcal{SS}.Recon(\{(\mathcal{T}_u, s_u)\}_{\mathcal{T}_u \in \mathcal{T}'}, t) \rightarrow s$ takes as input the threshold t and the shares corresponding to a subset $\mathcal{T}' \subseteq \mathcal{T}$ such that $|\mathcal{T}'| \geq t$, and outputs an element s.

Key Agreement. A key agreement protocol $\mathcal{KA}$ consists of a tuple of algorithms ($\mathcal{KA}.ParamGen, \mathcal{KA}.KeyGen, \mathcal{KA}.Agree$). The algorithm $\mathcal{KA}.ParamGen(\lambda) \rightarrow param$ produces some public parameters. $\mathcal{KA}.KeyGen(param) \rightarrow (pk_u, sk_u)$ allows the u-th participant to generate a private-public key pair. $\mathcal{KA}.Agree$ $(sk_u, pk_v) \rightarrow s_{u,v}$ allows any participant u to combine the private key sk_u with the public key pk_v for any v, to generate a shared key $s_{u,v}$ between u and v.

The specific key agreement scheme that will be used here is Diffie-Hellman key agreement which is composed with a hash function. More specifically, $\mathcal{KA}.ParamGen(\lambda) \rightarrow param = (\mathcal{G}', g, q, H)$ samples group $\mathcal{G}'$ of prime order q, along with a generator g, and a hash function H; $\mathcal{KA}.KeyGen((\mathcal{G}', g, q, H) \rightarrow (x, g^x)$ samples a random $x \in \mathbb{Z}_q$ as the secret key sk_u, and g^x as the public key pk_u; and $\mathcal{KA}.Agree$ $(x, g^x) \rightarrow s_{u,v}$ outputs $s_{u,v} = H((g^{x_v})x_u)$.

The scheme is correct if for any key pairs generated by trainers u and v, there is $\mathcal{KA}.Agree\,(sk_u, pk_v) = \mathcal{KA}.Agree\,(sk_v, pk_u)$. In the honest-but-curious model, for achieve the security requirement, it should meet that for any adversary who is given two honestly generated public keys pk_u and pk_v without the corresponding secret keys sk_u and sk_v, the shared secret $s_{u,v}$ computed from those keys is indistinguishable from a uniformly random string.

Authenticated Encryption. A authenticated encryption scheme $\mathcal{AE}$ guarantees confidentiality and integrity for messages exchanged between two parties. It consists of a tuple of algorithms $(\mathcal{AE}.\,KeyGen, \mathcal{AE}.Enc, \mathcal{AE}.Dec)$. Specially, the decryption algorithm $\mathcal{AE}.Dec$ that takes as input a ciphertext with a key and outputs either the original plaintext or $\perp$.

Pseudorandom Generator. Here, it is required a secure Pseudorandom Generator (PRG) PRG that takes in a uniformly random seed of some fixed length. The dimension of the output is m, which is accordance with the input of the protocol. The output space of PRG is $[0, R)^m$. Security for a Pseudorandom Generator guarantees that the output of a uniformly random seed is computationally indistinguishable from a uniformly sampled element of the output space, as long as the seed is hidden from the distinguisher.

Signature Scheme. A public-key signature scheme $\mathcal{SIG}$ consists of a tuple of algorithms $(\mathcal{SIG}.\,KeyGen, \mathcal{KA}.Sign, \mathcal{KA}.Ver)$. The key generation algorithm $\mathcal{SIG}.KeyGen(param) \rightarrow (pk_u, sk_u)$ takes as input the security parameter and outputs a key pair. The signing algorithm $\mathcal{SIG}.Sign(sk_u, m) \rightarrow \sigma$ takes as input a secret key and a message and outputs a signature σ. The verification algorithm $\mathcal{SIG}.Ver(pk_u, m, \sigma) \rightarrow b$ takes as input a public key, a message, and a signature, and returns a bit b that indicates whether σ is valid.

4.2.2 Technical Intuition

To handle trainer's drop while enabling the server to aggregate each $\mathbf{x}_u$, this work introduce a *double-masking* blinding structure to keep $\mathbf{x}_u$ private in the aggregation protocol.

Double Masking. Firstly, each trainer $\mathcal{T}_u$ runs the key agreement algorithm $\mathcal{KA}$ according to other owners to obtain several seed vectors $\mathbf{s}_{u,v}(1 \leq u \leq N, v \neq u)$. The first mask is a sum of some $PRG(\mathbf{s}_{u,v})$. $PRG(\mathbf{s}_{u,v})$ is the output of a pseudorandom generator (PRG), and has the same size as $\mathbf{x}_u$. Then, each trainer $\mathcal{T}_u$ chooses an additional random seed $\mathbf{b}_u$. Using the secret sharing scheme $\mathcal{SS}$, the trainer generates and distributes shares of $\mathbf{b}_u$ to others. To blind $\mathbf{x}_u$, $\mathcal{T}_u$ add two masks to it as follows.

$$\mathbf{y}_u \leftarrow \mathbf{x}_u + PRG(\mathbf{b}_u) + \left(\sum_{v=u+1}^{N} PRG(\mathbf{s}_{u,v}) - \sum_{v=1}^{u-1} PRG(\mathbf{s}_{u,v}) \right). \tag{4.1}$$

When the untrusted server recover the aggregation result of vectors, it must make an explicit choice with respect to each $\mathcal{T}_u$. In more details, from each non-dropped trainer $\mathcal{T}_u$, the server can request either a share of the common secret $\mathbf{s}_{u,v}$ associated with $\mathcal{T}_u$ or a share of the $\mathbf{b}_u$ for $\mathcal{T}_u$. Then, an honest $\mathcal{T}_v$ will never reveal the two shares for the same trainer. After gathering at least t shares of $\mathbf{s}_{u,v}$ for all dropped trainers and $\mathbf{b}_u$ for all non-dropped trainers respectively, the server can subtract off all the masks to recover the sum.

4.2.3   Secure Protocol

The secure protocol is run between a server $\mathcal{S}$ and N trainers, and consists of five steps. Each trainer $\mathcal{T}_u \in \mathcal{T}$ holds a private vector $\mathbf{x}_u$ of dimension m. The server can communicate with the trainers via secure channels. Trainers can drop out of the protocol at any time, and the server is still able to produce a correct output as long as the number of non-dropped trainers can meet the threshold t.

At the beginning, an honest system administrator generates public parameters for the protocol. These parameters include the security parameter k, a threshold value t, a field $\mathbb{F}$, parameters pp of a key agreement scheme $\mathcal{KA}$. The administrator issues each $\mathcal{T}_u$ a signature secret key sk_u^1 and publish all verification keys $\{pk_u^1\}_{u \in [N]}$. The procedures of the protocol are summarized as follows.

- *Initialization.*
  Each non-dropped trainer $\mathcal{T}_u$:

  - generates key pairs $(pk_u^2,\ sk_u^2) \leftarrow \mathcal{KA}.KeyGen(pp)$, $(pk_u^3,\ sk_u^3) \leftarrow \mathcal{KA}.KeyGen(pp)$, and a signature $\sigma_u \leftarrow \mathcal{SIG}.Sign(sk_u^1, pk_u^2||pk_u^3)$;
  - sends $pk_u^2||pk_u^3||\sigma_u$ to $\mathcal{S}$ via the secure channel.

  Server $\mathcal{S}$:

  - collects messages from a trainer set $\mathcal{T}'$ that contains at least t non-dropped trainers;
  - broadcasts each $(v,\ pk_v^2,\ pk_v^3,\ \sigma_v)$ to each $\mathcal{T}_v \in \mathcal{T}'$.

- *Key Sharing.*
  Each non-dropped trainer $\mathcal{T}_u$:

  - validates $\mathcal{SIG}.Ver(pk_v^1,\ pk_v^2||pk_v^3,\ \sigma_v) = 1$ for each $\mathcal{T}_v \in \mathcal{T}'$;
  - chooses a random seed $\mathbf{b}_u \in \mathbb{F}$ for the PRG $PRG(\cdot)$;
  - generates t-out-of-$|\mathcal{T}'|$ shares of sk_u^3 as $\{(\mathcal{T}_v, sk_{u,v}^3)\}_{\mathcal{T}_v \in \mathcal{T}'} \rightarrow \mathcal{SS}.Share$ $(sk_u^3, t, \mathcal{T}')$;
  - generates t-out-of-$|\mathcal{T}'|$ shares of $\mathbf{b}_u$ as $\{(\mathcal{T}_v, \mathbf{b}_{u,v})\}_{\mathcal{T}_v \in \mathcal{T}'} \rightarrow \mathcal{SS}.Share$ $(\mathbf{b}_{u,t}, \mathcal{T}')$;
  - for each other trainer $\mathcal{T}_v \in \mathcal{T}' \backslash \mathcal{T}_u$, computes the encryption $e_{u,v} \leftarrow \mathcal{AE}.Enc$ $(\mathcal{KA}.Agree(pk_u^2, pk_v^2), u||v||sk_{u,v}^3||\mathbf{b}_{u,v})$ and sends it to $\mathcal{S}$.

Server $\mathcal{S}$:

- collects ciphertexts from a trainer set $\mathcal{T}''$ that contains at least t non-dropped trainers;
- sends $\{e_{u,v}\}_{\mathcal{T}_v \in \mathcal{T}''}$ to each $\mathcal{T}_u \in \mathcal{T}''$.

- *Input Masking.*
  Each non-dropped trainer $\mathcal{T}_u$:

  - for each other trainer $\mathcal{T}_v \in \mathcal{T}'' \backslash \mathcal{T}_u$, computes $\mathcal{KA}.Agree(pk_u^3, pk_v^3)$ and expands the result to $\mathbf{s}_{u,v}$;
  - masks $\mathbf{x}_u$ using Eq. 4.1 and sends the result $\mathbf{y}_u$ to $\mathcal{S}$.

Server $\mathcal{S}$:

- collects $\mathbf{y}_u$ from a trainer set $\mathcal{T}'''$ that contains at least t non-dropped trainers and sends the list of $\mathcal{T}'''$ to each $\mathcal{T}_u \in \mathcal{T}'''$.

- *Consistency Checking.*
  Each non-dropped trainer $\mathcal{T}_u$:

  - generate a signature $\sigma_u' \leftarrow \mathcal{SIG}.Sign(sk_u^1, \mathcal{T}''')$.

Server $\mathcal{S}$:

- collects σ_u' from a trainer set $\mathcal{T}''''$ that contains at least t non-dropped trainers;
- sends $\{(v, \sigma_v')\}_{\mathcal{T}_v \in \mathcal{T}''''}$ to each $\mathcal{T}_u \in \mathcal{T}''''$.

- *Unmasking.*
  Each non-dropped trainer $\mathcal{T}_u$:

  - validates $\mathcal{SIG}.Ver(pk_v^1, \mathcal{T}''', \sigma_v') = 1$ for each $\mathcal{T}_v \in \mathcal{T}''''$;
  - for each other trainer $\mathcal{T}_v \in \mathcal{T}'' \backslash \mathcal{T}_u$, performs the decryption to obtain $v'||u'||sk_{v,u}^3 ||\mathbf{b}_{v,u} \leftarrow \mathcal{AE}.Dec(\mathcal{KA}.Agree(pk_u^2, pk_v^2), e_{v,u})$ and validates $u = u'$ and $v = v'$;
  - sends $\{sk_{v,u}^3\}_{\mathcal{T}_v \in \mathcal{T}'' \backslash \mathcal{T}'''}$ and $\{\mathbf{b}_{v,u}\}_{\mathcal{T}_v \in \mathcal{T}'''}$ to $\mathcal{S}$.

Server $\mathcal{S}$:

- collects responses from a trainer set $\mathcal{T}'''''$ that contains at least t non-dropped trainers;
- for each $\mathcal{T}_u \in \mathcal{T}'' \backslash \mathcal{T}'''$, reconstructs $sk_u^3 \leftarrow \mathcal{SS}.Recon(\{sk_{u,v}^3\}_{\mathcal{T}_v \in \mathcal{T}'''''}, t)$ and uses it to compute $\mathbf{s}_{v,u}$ for each $\mathcal{T}_v \in \mathcal{T}'''$;
- for each $\mathcal{T}_u \in \mathcal{T}'''$, reconstructs $\mathbf{b}_u \leftarrow \mathcal{SS}.Recon(\{\mathbf{b}_{u,v}\}_{\mathcal{T}_v \in \mathcal{T}'''''}, t)$;
- computes and outputs the aggregation result $\mathbf{z} \leftarrow \sum_{\mathcal{T}_u \in \mathcal{T}'''} \mathbf{y}_u - \sum_{\mathcal{T}_u \in \mathcal{T}'''} PRG(\mathbf{b}_u) + \sum_{\mathcal{T}_u \in \mathcal{T}''', \mathcal{T}_v \in \mathcal{T}'' \backslash \mathcal{T}'''} PRG(\mathbf{s}_{v,u})$.

References

Bonawitz K, Ivanov V, Kreuter B, Marcedone A, McMahan HB, Patel S, Ramage D, Segal A, Seth
    K (2017) Practical secure aggregation for privacy-preserving machine learning. In: Proceedings
    of the 2017 ACM SIGSAC conference on computer and communications security. ACM, pp
    1175–1191
Shokri R, Shmatikov V (2015) Privacy-preserving deep learning. In: Proceedings of the 22nd ACM
    SIGSAC conference on computer and communications security. ACM, pp 1310–1321

Chapter 5
Learning with Differential Privacy

5.1 Differential Privacy

Considering the internal representations in machine learning models may potentially imply some of training data, an adversary may launch attacks to extract parts of training data from a trained model (e.g., the "black-box" model-inversion attack Fredrikson et al. 2015). Thus, the privacy guarantees if datasets contain correlated inputs is also important.

5.1.1 Definition

Differential privacy Dwork and Roth (2014) addresses the paradox of learning nothing about an individual while learning useful information about a population. It is a promise made by a data holder or curator. Differentially private database mechanisms can make confidential data widely available for accurate data analysis, without resorting to data clean rooms, data usage agreements, data protection plans, or restricted views.

Differential privacy is a definition of privacy tailored to the problem of privacy-preserving data analysis, such as machine learning tasks. For a given computational task T and a given value of ϵ there will be many differentially private algorithms for achieving T in an ϵ-differentially private manner. Some will have better accuracy than others. When ϵ is small, finding a highly accurate ϵ-differentially private algorithm for T can be difficult, much as finding a numerically stable algorithm for a specific computational task can require effort.

Assume the existence of a totally trusted administrator who holds the data of individuals in a database D which typically contains n rows of data. It is intuitive that each row contains the data of a single individual. There is supposed to be a

© The Author(s), under exclusive license to Springer Nature Singapore Pte Ltd. 2022
J. Li et al., *Privacy-Preserving Machine Learning*, SpringerBriefs on Cyber Security Systems and Networks, https://doi.org/10.1007/978-981-16-9139-3_5

privacy goal in which the privacy of every individual row is protected at the same time, even if statistical analysis of the whole database is allowed.

A query on the database can be seen as a function. An interactive or on-line database service permits the user to make queries adaptively. That is, the observed responses to previous queries can be used to decide which query should be posed next. Using the cryptographic techniques for secure multi-party protocols, the trusted administrator can be replaced by a protocol run by the set of individuals. If all the queries are known in advance, the non-interactive database will give the best accuracy, since knowing the structure of the queries can help to correlate noises. Alternatively, if there is no information about the queries in advance, the non-interactive database must provide answers to all possible queries. To ensure privacy, the accuracy will be certainly dedicated with the number of questions queried, since it is infeasible to provide accurate answers to all possible questions.

Regardless of an adversary's background knowledge and available computational power, differential Privacy provides a formal and quantifiable privacy guarantee. Dwork (2014) shows that the differential Privacy is actually a condition on the data release mechanism but not on the dataset. A randomized algorithm is considered to be differentially private if for any pair of neighbouring inputs, the probability of generating the same output is within a small multiple of each other for the entire output space. This means that for any two datasets which are very close, a differentially private algorithm will behave approximately the same on both datasets. This notion provides sufficient privacy protection for users regardless of the prior knowledges possessed by the adversaries.

(ϵ, δ)—**Differential Privacy**. A randomization algorithm $\mathcal{M}$ satisfies (ϵ, δ)—*differential privacy* if for any two neighbouring datasets D_1 and D_2, and any output $D \subseteq Range(\mathcal{M})$, we have $Pr[\mathcal{M}(D_1) \in D] \leq e^{\epsilon} Pr[\mathcal{M}(D_2) \in D] + \delta$.

Then, we talk about how to implement (ϵ, δ)—Differential Privacy on a given dataset.

5.1.2 Privacy Mechanism

Dwork and Roth (2014) gives a definition of the privacy mechanism. A privacy mechanism is an algorithm which is shown as a randomization algorithm $\mathcal{M}$ above. It takes as input a database, a universe $\mathcal{X}$ of data types, and optionally a set of queries and probabilistically outputs an string. It is supposed that the output string can be decoded to produce relatively accurate answers to the queries, if there is quite an answer. If no queries are presented then the non-interactive case works. In this case, we hope that the output string can be interpreted to provide answers to future queries. In some cases, the output string is required to be a synthetic database, which can be seen as a multi-attribute set drawn from the universe $\mathcal{X}$ of possible database rows. The decoding method in this case is to carry out the query on the synthetic database and then to apply some sort of simple transformation, such as multiplying by a scaling factor, to obtain an approximation to the true answer to the query.

Given any function $f : \mathbb{N}^{|\mathcal{X}|} \to \mathbb{R}^k$, a private mechanism is defined as

$$\mathcal{M}_L(x, f(\cdot), \epsilon) = f(x) + (Y_1, ... Y_k). \tag{5.1}$$

Numeric queries, are one of the most fundamental types of dataset queries. These queries map databases to some real numbers. One of the important parameters that will determine just how accurately such queries can be given is their sensitivity.

Sensitivity. For a function $f : D \to \mathbb{R}^d$ over the input datasets, the sensitivity of f is $\Delta f = max||f(D_1) - f(D_2)||$ for any two neighbouring datasets D_1 and D_2.

The sensitivity of a function f captures the magnitude by which a single individual's data can change the function f in the worst case. Therefore, the sensitivity of a function gives an upper bound on how much we must perturb its output to preserve privacy. One noise distribution naturally lends itself to differential privacy. The lower sensitivity queries with, the better tolerate the data modifications from added noise is.

The Laplace Distribution. The Laplace Distribution (centred at 0) with scale b is the distribution with probability density function:

$$Lap(x|b) = \frac{1}{2b} \exp(-\frac{|x|}{b}).$$

Laplace distribution with scale b can be simply denoted as $Lap(b)$. Sometimes abuse notation and write $Lap(b)$ simply to denote a random variable $X \sim Lap(b)$. The Laplace distribution is a symmetric version of the exponential distribution. The Laplace mechanism will simply compute f, and perturb each coordinate with noise drawn from the Laplace distribution. The scale of the noise will be calibrated to the sensitivity of f.

The Laplace Mechanism. Given any function $f : \mathbb{N}^{|\mathcal{X}|} \to \mathbb{R}^k$, the Laplace mechanism is defined as:

$$\mathcal{M}_L(x, f(\cdot), \epsilon) = f(x) + (Y_1, ... Y_k)$$

where each noise Y_i is i.i.d. random variable drawn from $Lap(\Delta f/\epsilon)$. As proofed in Dwork and Roth (2014), some private mechanisms $(\epsilon, 0)$-differential privacy (or simply ϵ-differential privacy), which shows that releasing a function $f(\cdot)$ with the mechanism $\mathcal{M}_L$ can satisfy privacy.

Sometimes, the sensitivity is needed to be independent with the universe $\mathcal{X}$ of a database. Especially, some private mechanisms apply additive noises which follows from the basic composition theorem or advanced composition theorems, such that they can be implemented on queries several times. Moreover, when executing the mechanism, it is possible to keep track of the accumulated privacy loss and enforcing the applicable privacy policy. Thus, another privacy mechanism is introduced.

Gaussian Mechanism. If each noise Y_i is i.i.d. random variable drawn from $\mathcal{N}(0, \Delta f^2 \cdot \sigma)$ (i.e., *Normal Distribution* centred at 0 with standard deviation Δf), $\delta > \frac{4}{5} e^{-\frac{(\epsilon\sigma)^2}{2}}$, and $\epsilon < 1$, $\mathcal{M}_L$ achieves (ϵ, δ)-differential privacy.

As proofed in Dwork and Roth (2014), some private mechanisms (ϵ, δ)-differential privacy, which shows that releasing a function $f(\cdot)$ with the mechanism $\mathcal{M}_L$ can satisfy privacy.

5.2   Deep Learning with Differential Privacy

Abadi et al. (2016) proposed an unsupervised locally training scheme for the deep neural network model with differential privacy and a refined analysis of privacy costs. Instead of the error function $E(\cdot)$, this work defines a loss function $L(\cdot)$ that represents the penalty for mismatching training data. The reason is that training consists in finding neural network parameters $\mathbf{w}$ that yields an acceptably small loss similar to the small error. $L(\mathbf{w})$ with parameters $\mathbf{w}$ indicates the average of the loss over the training set D. Thus, the loss function is

$$L(\mathbf{w}) = \frac{1}{N} \sum_{\mathbf{x} \in D} L(\mathbf{w}, \mathbf{x}). \tag{5.2}$$

This work also chooses the Stochastic gradient descent (SGD) as the update rule of parameters $\mathbf{w}$. During the local training, the minimization of $L(\mathbf{w})$ is often done by the mini-batch SGD. That is, at each step, the trainer forms a batch B of random training data and computes

$$\mathbf{g}_B = \frac{1}{|B|} \sum_{\mathbf{x} \in B} \nabla_{\mathbf{w}} L(\mathbf{w}, \mathbf{x}). \tag{5.3}$$

as an estimation to the gradient $\nabla_{\mathbf{w}} L(\mathbf{w})$. Then, $\mathbf{w}$ will be updated following the gradient direction $-\mathbf{g}_B$ towards a local minimum.

5.2.1   Differentially Private SGD Algorithm

Treating the learning process as a black box, simply protecting the privacy in training data by working only on the final trained deep neural network model may add overly conservative noises that are selected according to the worst-case analysis, which would destroy the utility of the model. The main local training approach is the differentially private SGD algorithm, where private mechanisms are performed on the gradients.

Algorithm 1 describes the basic method for training a deep neural network model with parameters $\mathbf{w}$ by minimizing the empirical loss $L(\mathbf{w})$. At each epoch of the SGD, the trainer firstly computes the gradient $\nabla_{\mathbf{w}} L(\mathbf{w})$ for a random subset $B \subseteq D$ where D is the whole dataset.

To achieve the differential privacy that can be proved, it requires bounding the influence of each individual training data on a noised gradient. Since there is no a priori bound on the size of the gradients, the trainer then has to clip each gradient in l_2 norm. That is, the gradient vector $\mathbf{g}$ is replaced by according to a clipping threshold C. If $||\mathbf{g}||_2 > C$, $\mathbf{g}$ gets scaled down to be of norm C, otherwise it is preserved.

At the end of the epoch, the trainer computes the average, performs the private mechanism, and adjusts $\mathbf{w}$ by the opposite direction of this average noises gradient.

Algorithm 1 Differentially Private SGD Algorithm.

Input:
  training dataset $D = \{\mathbf{x}_1, ..., \mathbf{x}_N\}$ that contains N training instances, loss function $L(\mathbf{w})$, the number of training epochs T, learning rate η, noise scale σ, and gradient norm bound C

Output:
  final parameters of trained deep neural network model $\mathbf{w}_T$

1: Randomly initialize the network model $\mathbf{w}_0$;
2: **for** each $t \in [T]$ **do**
3:     Sample a random subset $B_t \subseteq D$ with sampling probability $|B|/N$;
4:     **for** each $\mathbf{x} \in B_t$ **do**
5:         // Compute gradient
6:         Compute $\mathbf{g}_t(\mathbf{x}) \leftarrow \nabla_{\mathbf{w}_t} L(\mathbf{w}_t, \mathbf{x})$;
7:         // Clip gradient
8:         Compute $\bar{\mathbf{g}}_t(\mathbf{x}) \leftarrow \mathbf{g}_t(\mathbf{x}) / \max(1, \frac{||\mathbf{g}_t(\mathbf{x})||}{C})$;
9:     **end for**
10:     // Add noise
11:     Compute $\tilde{\mathbf{g}}_t \leftarrow \frac{1}{|B_t|} \sum_{\mathbf{x} \in B_t} (\bar{\mathbf{g}}_t(\mathbf{x}) + \mathcal{N}(0, \sigma^2 C^2 \mathbf{I}))$;
12:     // Adjust
13:     $\mathbf{w}_{t+1} \leftarrow \mathbf{w}_t - \eta \tilde{\mathbf{g}}_t$;
14: **end for**
15: **return** $\mathbf{w}_T$.

5.2.2 Privacy Account

The pseudo-code in Algorithm 1 treats all the parameters as a single input θ of the loss function $L(\theta)$. In a neural network with the multi-layer perception structure, different layers can have different clipping thresholds C and noise scales if each layer is considered separately. Such a clipping and noise parameters may vary with the number of training steps t.

As the non-privacy-preserving SGD algorithm, the algorithm estimates the gradient of L by computing the averaged gradient of the loss function on a mini-batch of training data. The variance of the unbiased estimator provided by this average will quickly decrease with the size of the mini-batch. For improving performance and avoiding overfitting, the batch size is set much smaller than the lot's size L. At each epochs, the learning is performed in a mini-batch and the noised results are related the mini-batch. As is common in the literature, the running time of a training algorithm should be normalized by expressing it as the number of epochs, where each epoch is the expected number of batches required to process N examples. In the notation above, an epoch consists of N/L lots.

For making a algorithm differentially private, an important issue is computing the global privacy cost of the training function. The aggregation property of Gauss mechanism inspire the work to implement an accountant procedure that computes the privacy cost at each access to the training data, and accumulates this cost as the training progresses. The gradients at multiple layers are computed at each epochs, and the accountant accumulates the cost that corresponds to the gradients round by round.

As described in Abadi et al. (2016), for the Gaussian noise, if θ is chosen in the algorithm to be $\sqrt{2log(\frac{1.25}{\theta})}/\epsilon$, then by standard arguments each step is (ϵ, δ)-differentially private with respect to the lot. Since the lot itself is randomly selected from the database, the privacy amplification theorem implies that each step is $(q\epsilon, q\delta)$-differentially private with respect to the full database where $q = L/N$ is the sampling ratio per lot. This scheme invents a stronger accounting method, which is the moments accountant. It is allowed to prove that the algorithm is $(O(q\epsilon\sqrt{T}), \delta)$-differentially private for appropriately chosen settings of the noise scale and the clipping threshold. A tighter bound is achieved by saving a $\sqrt{log(\frac{1}{\delta})}$ factor in the ϵ part and a Tq factor in the δ part. Since it is expected δ to be small and $T >> 1/q$, the saving provided by the bound is significant in the implementation.

5.3   Distributed Deep Learning with Differential Privacy

As we discuss in previous chapter, Shokri and Shmatikov (2015) proposed a distributed system that enables multiple trainers to jointly learn an accurate deep learning network model without revealing their training datasets. In this system, author also designed an approach to achieve (ϵ, δ)-differential privacy in a trained deep learning model.

5.3.1   Private Algorithm

In this work, the function f is defined to compute parameter gradients and select which of them to share with other trainers. If the gradients are seen as sensitive information, their privacy should be protected. There are two sources of potential leakage: how they are selected for sharing and the their actual values. As shown in previous chapter, randomly selecting a small subset of gradients whose values are above a threshold can mitigate both types of leakage. Moreover, a consistent differentially private mechanism can be used to share perturbed values of the selected gradients.

The private mechanism is to release the responses to queries whose value is above a publicly threshold. Let ϵ be the each participant's total privacy budget and c be the total number of gradients uploaded (i.e., $c = \theta_u|\Delta\mathbf{w}|$) at each epoch. The budget for each potential uploaded gradient includes two parts. The first will be spent on checking whether a randomly chosen parameter j's gradient $\Delta w_j^{(i)}$ is above the threshold τ. If so, the second will be spent on making the actually uploaded gradient to achieve differential privacy. This work adopts the Laplace mechanism to add noises before uploading gradients according to the predetermined privacy budgets. The Laplace noise depends on the privacy budget along with the gradient's sensitivity Δf for the query f.

This work uses $8/9$ of ϵ/c to feed the selection in which one part is spent on random noise r_w and the other is spent on random noise r_τ. Alternatively, the remaining $1/9$ is devoted to the released values.

Since r_τ is not re-generated even the threshold check is failed, all uploaded gradients are differentially private and other gradients do not cost any extra budget. The private algorithm in Shokri and Shmatikov (2015) is described as follows.

- Let ϵ be the total privacy budget for one epoch of participant i running DSSGD, and let Δf be the sensitivity of each gradient.
- Let $c = \theta_u|\Delta\mathbf{w}|$ be the maximum number of gradients that can be uploaded in one epoch.
- Let $\epsilon_1 = \frac{8}{9}\epsilon$, $\epsilon_2 = \frac{2}{9}\epsilon$.
- Let $\sigma(x) = \frac{2c\delta f}{x}$.

1 Generate fresh random noise $r_\tau \sim Lap(\sigma(\epsilon_1))$.
2 Randomly select a gradient $\Delta w_j^{(i)}$.
3 Generate fresh random noise $r_w \sim Lap(2\sigma(\epsilon_1))$.
4 If $|bound(\Delta w_j^{(i)}, \gamma)| + r_w \geq r + r_\tau$, then

    (a) Generate fresh random noise $r_w' \sim Lap(\sigma(\epsilon_2))$;
    (b) Upload $bound(\Delta w_j^{(i)} + r_w', \gamma) + r_w$ to the parameter server;
    (c) Charge ϵ/c to the privacy budget;
    (d) If number of uploaded gradients is equal to c, then Halt Else Goto Step 1.
     5 Else Goto Step 2.

5.3.2 Estimating Sensitivity

As shown at the start of this chapter, for Laplace mechanism, the sensitivity of a query (function) determines how much Laplace noise needs to be added to the response to achieve differential privacy. The sensitivity of f is

$$\Delta f = max||f(D) - f(D')||.$$

Since SGD is not a linear function, estimating the it's sensitivity of SGD is a challenge. The work Shokri and Shmatikov (2015) proposes an approach for making the function's output stays within fixed. The modified function is input-independent such that its bound can be used to estimate sensitivity. That is, the bound function enforces a $[-\gamma, \gamma]$ range on gradient values shared with other trainers. In fact, this approach can guarantee differential privacy even though it may degrade accuracy.

As discussed in previous section, increasing randomness of values and limiting its range of values can help to avoid overfitting so that improve the training process as well as regularization techniques. In SGD, gradient values indicate the direction of moves while truncating into the $[-\gamma, \gamma]$ range limits the magnitude of moves. Therefore, small values of γ not only implies smaller sensitivity and thus smaller noise and higher accuracy, but also influences the learning rate of the algorithm. Moreover, since gradients of each trainer are aggregated, the gradient descent algorithm can still achieve the local optima.

Since the parameter γ is independent of the training data, the output does not leak any sensitive information. The sensitivity of the function is estimated as 2γ and the uploaded gradients are truncated into the $[-\gamma, \gamma]$ range. As a known regularization technique, the truncation is to actually bound the norm of all gradients and thus result in higher accuracy. This helps to achieve an acceptable trade-off between the detrimental effect of large noise values and privacy preserving during training.

References

Abadi M, Chu A, Goodfellow I, McMahan HB, Mironov I, Talwar I, Zhang L (2016) Deep learning with differential privacy. In: Proceedings of the 2016 ACM SIGSAC conference on computer and communications security. ACM, pp 308–318

Dwork C, Roth A et al (2014) The algorithmic foundations of differential privacy,. Found Trends® Theor Comput Sci 9(3–4):211–407

Fredrikson M, Jha S, Ristenpart T (2015) Model inversion attacks that exploit confidence information and basic countermeasures. In: Proceedings of the 22nd ACM SIGSAC conference on computer and communications security. ACM, pp 1322–1333

Shokri R, Shmatikov V (2015) Privacy-preserving deep learning. In: Proceedings of the 22nd ACM SIGSAC conference on computer and communications security, ACM, pp 1310–1321

Chapter 6
Applications—Privacy-Preserving Image Processing

6.1 Machine Learning Image Processing for Privacy Protection

In the following chapters, we introduce the fusion of machine learning, image processing and security as an emerging paradigm to protect the privacy of clients and classifier. If the model of machine learning image processing is supervised, then it can be divided into two phases: (i) the training phase during which the algorithm learns a model φ from a data set of labeled samples, and (ii) the classification phase that runs a classifier C over a previously unseen feature vector y, using the trained model φ to output a prediction $C(y; \varphi)$.

In the Fig. 6.1, we show the generally model of machine learning image processing. The database and model are held by the cloud server, and the input and classification result are supported by the client. Hence, if we want to protect the confidential of private data in this model, we need to focus on three points: a secure training phase that protects the database and model φ; a secure classification phase that protects the classification parameters during classifying; or both a secure training and classification which not only protects the confidential of input data (provided by both cloud server and client) and model φ but the parameter of the classifier C. Due to the advent of big data and cloud computing, machine learning as a service, lots of works about the design of outsourcing image processing based on machine learning have been proposed.

Machine learning image processing have aroused privacy concerns in recent years, which applied to conduct client-cloud related data. However, if these related data contain the client's privacy-sensitive information, then machine learning image processing is operated on cloud server can be abused, make economic loss or personal safety.

In the line of privacy-preserving training and classification phase, they are some works have been done. Graepel et al. (2012) consider several machine learning classifiers (e.g. LM, FLD) based on encrypted data using somewhat homomorphic encryption (SHE) cryptosystem in a passive model. However, in their scheme, performing

J. Li et al., *Privacy-Preserving Machine Learning*, SpringerBriefs on Cyber Security Systems and Networks, https://doi.org/10.1007/978-981-16-9139-3_6

Fig. 6.1 Framework of model

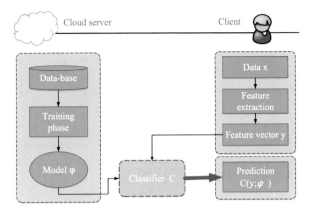

the final comparison with fully homomorphic encryption is inefficient. To achieve the passive security, they assume the two servers without colluding.

6.2 Feature Extraction Methods of Machine Learning Image Processing

Recently, scale invariant feature transform (SIFT) is a hot issues in computer vision to extract and describe local features in image, there are four major computational stages to generate the set of image features. Here, we list mainly two phase as follows:

Scale-space Extreme Detection. In order to identify potential interest points that are invariant to scale and orientation, we use a Difference-of-Gaussian (DoG) function to convolve with image at multiple scales. Firstly, to construct scale space, while the element of this space is an image $f(x, y)$ convolved with Gaussian function $G(x, y, \sigma)$:

$$L(x, y, \sigma) = G(x, y, \sigma) * f(x, y)$$
$$= \sum_{u,v} G(u, v, \sigma) f(x - u, y - v), \qquad (6.1)$$

where $G(x, y, \sigma) = \frac{1}{2\pi\sigma^2} e^{-\frac{x^2+y^2}{2\sigma^2}}$, σ notes variance and the scale size. If the size is bigger, the Guassian-blurred image in scale space is coarser, contrast, is more finer.

Secondly, we need to construct DoG space that detect candidate keypoint locations efficiently. The element of DoG space can be computed like this:

$$D\_f(x, y, \sigma) = (G(x, y, k\sigma) - G(x, y, \sigma)) * f(x, y) = L(x, y, k\sigma) - L(x, y, \sigma), \quad (6.2)$$

which means the difference of nearby scales convolved with an image $f(x, y)$.

Thirdly, to detect the potential feature points in DoG space, we need to detect the local maxima and minima of $D\_f(x, y, \sigma)$, each sample point is compared to its eight neighbors in the current image and nine neighbors in the scale above and below. If this sample point is the extreme among all neighbors, then it is selected as a keypoint.

Keypoint Descriptor. Based on the gradient directions of the local image, each keypoint location is assigned orientations. A 128-dimensional SIFT keypoint descriptor is established for the 16-16 region, which is further divided into sixteen 4-4 blocks, around a feature point. The descriptor is represented as a vector containing the values of all assigned orientation histogram entries. That is for each scale image sample $L(x, y)$ at this scale, the gradient magnitude $m(x, y)$ and orientation $\theta(x, y)$ is precomputed using pixel differences:

$$m(x, y) = \sqrt{(L(x + 1, y) - L(x - 1, y))^2 + (L(x, y + 1) - L(x, y - 1))^2},$$
$$\theta(x, y) = tan^{-1}(((L(x, y + 1) - L(x, y - 1))/(L(x + 1, y) - L(x - 1, y)))$$
$$(6.3)$$

where $L(x + 1, y) - L(x - 1, y)$ and $L(x, y + 1) - L(x, y - 1))$ is the gradient magnitude at X-axis and Y-axis, respectively. The magnitude and direction computations for the gradient are done for every pixel (x, y) in a neighboring region around the keypoint in the Gaussian-blurred image. After the computations of the magnitude $m(x, y)$ and orientation $\theta(x, y)$, an orientation histogram with 36 bins is formed, with each bin covering $10°$. Each image sample point in the neighboring window added to a histogram bin is weighted by its gradient magnitude and a Gaussian-weighted circular window with a σ that is 1.5 times that of the keypoint's scale. The peaks in this histogram correspond to dominant orientations. Once the histogram is filled, the orientations corresponding to the highest peak and local peaks that are larger than 80% of the highest peaks are assigned to the keypoint.

6.3 Main Models of Machine Learning Image Processing for Privacy Protection

In this section we give several privacy-sensitive image processing models and describe how the privacy protection technology can be used in these models. Numerous algorithms of this models are based on eigenface technique.

6.3.1   Privacy-Preserving Face Recognition

In first privacy-sensitive image processing problem we consider is face recognition using PCA algorithm. The cryptosystem satisfy the additively homomorphic property. In this privacy-preserving face recognition model, Alice and Bob want to privately perform a face recognition protocol. Alice has a face image Γ (the query image), and Bob owns a image database containing a collection of face image from individuals. In this face recognition protocol, Alice want to know whether the query image Γ is stored in Bob' database without leaking any information about hers Γ, while Bob runs this protocol correctly and knows nothing about the query image and recognition result. During this protocol, Γ as private data for Alice, and the content of the database as private data for Bob that he is not willing to reveal. For example, the privacy-preserving face recognition model is illustrated in the following Fig. 6.2,

1. □ Alice outsources the encrypted query image $[\![\Gamma]\!]$ to Bob. Here the image Γ is a vector with length N. Hence, $[\![\Gamma]\!] = ([\![\Gamma_1]\!], [\![\Gamma_2]\!], \ldots, [\![\Gamma_N]\!])$.
2. □ After Bob received the outsourced image data $[\![\Gamma]\!]$, he projects the this encrypted data into the low-dimension face space spanned by the eigenfaces (basis) $u_1, u_2, \cdots, u_K$. Then he obtains a encrypted feature vector $[\![\overline{\Omega}]\!]$ of $[\![\Gamma]\!]$ in this face space. That is $[\![\overline{\Omega}]\!] = ([\![\overline{\omega}_1]\!], [\![\overline{\omega}_2]\!], \ldots, [\![\overline{\omega}_K]\!])^\tau$. Noting that the eigenfaces are obtained through a training algorithm when Bob builds the face database.
3. □ Assume that $\Omega_1, \Omega_2, \ldots, \Omega_K$ are the feature vector representation of training examples $\Theta_1, \Theta_2, \cdots, \Theta_M$ in face database. Bob computes the euclidean distance $[\![D_j]\!] := [\![\|\overline{\Omega} - \Omega_j\|_2]\!] (j = 1, 2, \cdots, M)$.

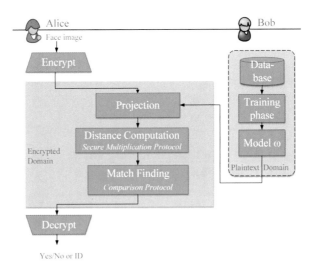

Fig. 6.2 Privacy-preserving face recognition

Table 6.1 Secure multiplication protocol

Secure multiplication protocol		
Alice		Bob
Holds public/private key pk_a/sk_a Decrypts $[\![z_{ji}]\!]$ and computes z_{ji}^2	$\xleftarrow{\forall i=1,2,\dots,n}$ $[\![z_{ji}]\!]$ $\xrightarrow{\forall i=1,2,\dots,n}$ $[\![z_{ji}^2]\!]$	Holds public key pk_a for all $i = 1, 2, \dots, K$ uniformly chooses $r_{ji} \xleftarrow{U} \mathbb{P}$ computes $[\![z_{ji}]\!] =$ $[\![x_{ji} + r_{ji}]\!] = [\![x_{ji}]\!] \odot [\![r_{ji}]\!]$ computes $[\![x_{ji}^2]\!] =$ $[\![z_{ji}^2 - (2x_{ji} \cdot r_{ji} + r_{ji}^2)]\!]$ $= [\![z_{ji}^2]\!] \odot [\![x_{ji}]\!]^{-2r_{ji}} \odot [\![r_{ji}^2]\!]^{-1}$ computes $[\![\sum_{i=1}^{K} x_{ji}^2]\!] = \prod_{i=1}^{K}[\![x_{ji}^2]\!]$

4. ☐ Bob computes the minimum euclidean distance between $[\![D_1]\!]$, $[\![D_2]\!]$, ..., $[\![D_M]\!]$ and determines whether the minimum euclidean distance is smaller than a threshold T or not.
5. ☐ Bob sends the comparison result of the recognition process to Alice.

In this privacy-preserving face recognition model, two key subprotocols: *Secure Multiplication Protocol* and *Secure Comparison Protocol* should execute. Assume that the euclidean distance $D(\overline{\Omega}, \Omega_j)$ defined as

$$D_j = D(\overline{\Omega}, \Omega_j) := \|\overline{\Omega} - \Omega_j\|_2^2 = \sum_{i=1}^{K}(\overline{\omega}_i - \omega_{ji})^2 = \underbrace{\sum_{i=1}^{K}\overline{\omega}_i^2}_{S_1} + \underbrace{\sum_{i=1}^{K}(-2\overline{\omega}_i\omega_{ji})}_{S_2} + \underbrace{\sum_{i=1}^{K}\omega_{ji}^2}_{S_3}. \quad (6.4)$$

According to the homomorphic property, Eq. (6.4) can be computed as

$$[\![D_j]\!] = [\![D(\overline{\Omega}, \Omega_j)]\!] = [\![S_1]\!] \odot [\![S_2]\!] \odot [\![S_3]\!]. \quad (6.5)$$

Noting that Bob knows $\Omega_j = (\omega_{j1}, \omega_{j2}, \dots, \omega_{jK})$, then S_3 can be computed and $[\![S_3]\!]$ can be directly encrypted. Meanwhile, Bob knows the encrypted feature vectors $[\![\overline{\Omega}]\!] = ([\![\overline{\omega}_{j1}]\!], [\![\overline{\omega}_{j2}]\!], \cdots, [\![\overline{\omega}_{jK}]\!])$. The term $[\![(-2\omega_i)\overline{\omega}_{ji}]\!]$ can be seen as the $[\![\overline{\omega}_{ji}]\!]$ with the power of constant $(-2\omega_j)$ by using the homomorphic property. Thus, $[\![S_2]\!]$ can be computed by $[\![S_2]\!] = \prod_{i=1}^{K}[\![\overline{\omega}_{ji}]\!]^{-2\omega_i}$. To compute $[\![S_1]\!]$, Bob needs to execute a secure multiplication protocol Table 6.1 with Alice:

Later, Alice and Bob jointly compute a secure comparison protocol (Table 6.2) to achieve the minimum value D_{min} from the euclidean distance of $[\![D_1]\!]$, $[\![D_2]\!]$, $\cdots$, $[\![D_M]\!]$ and its index Id. If D_{min} is smaller than the given threshold value T known by Bob, then Bob gets the result ID.

Table 6.2 Secure comparison protocol

Alice		Bob
Comparison over encrypted data		
Holds public pk_a and private key sk_a, and ciphertext $[\![a]\!]$ Decrypts $[\![d]\!]$ and computes $d \bmod 2^l$	$\xleftarrow{\;[\![d]\!]\;}$ $\xrightarrow{\;[\![d \bmod 2^l]\!]\;}$	Holds public key pk_a, the bit length l of b and $\lambda \in \mathbb{N}$ computes $[\![z]\!] = [\![a + 2^l - b]\!] = [\![a]\!] \odot [\![2^l]\!] \odot [\![b]\!]^{-1}$; uniformly chooses $r \xleftarrow{U} (0, 2^{l+\lambda}) \cap \mathbb{Z}$; computes $[\![d]\!] = [\![z + r]\!] = [\![z]\!] \odot [\![r]\!]$; removes r from $[\![d \bmod 2^l]\!]$: $[\![z \bmod 2^l]\!] = [\![d \bmod 2^l]\!] \odot [\![r \bmod 2^l]\!]^{-1} \odot [\![t]\!]^{2^l}$

We can know $a < b$ iff the most important bit (denoted as z_l) of z is 0. Hence, z_l can be computed by $z_l = 2^{-1}(z - (z \bmod 2^l))$. If Bob had an encryption of $z \bmod 2^l$, the comparison result would be obtain. The representation $t2^l$ denotes a correction term with $t \in \{0, 1\}$ indicating whether $d \bmod 2^l$ is larger or smaller than $r \bmod 2^l$. To obtain the encrypted value $[\![t]\!]$, we refer to details on the Yao's millionaire as a subprotocol to comparison.

6.3.2 Privacy-Preserving Object Recognition

The second privacy-sensitive image processing problem we consider in more detail is object recognition using SIFT algorithm under the additively homomorphic cryptosystem. Figure 6.3 illustrates a framework of proposed privacy-preserving SIFT based method and applications. This model is similarities to the privacy-preserving face-recognition solution (Sect. 6.3.1). However, there are some differences between the two solutions:

1. Bob holds database and stores lots of feature descriptors over plaintext domain. There is no training phase to learn a classification model used to project the query image.
2. In *match finding* phase, the feature descriptor of the query image is compared with the database over encrypted domain.

The SIFT algorithm over encrypted data consists of the following steps:

1. ☐ To protect Alice' privacy, Alice encrypts query image $f(x, y)$ before outsourcing to Bob for image. That is sends $Enc(f(x, y)) = [\![f(x, y)]\!]$ to Bob.
2. ☐ Firstly, Bob computes the integer DoG filter $G_D(u, v, \sigma_{ji})$ as follows:

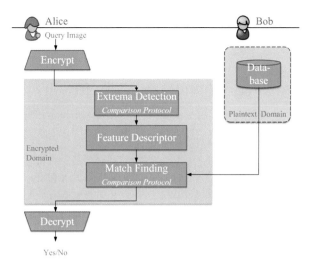

Fig. 6.3 Privacy-preserving SIFT

$$G_D(u, v, \sigma_{ji}) = \lfloor s(G(u, v, \sigma_j) - G(u, v, \sigma_i)) \rceil, \quad \forall u \text{ and } v \qquad (6.6)$$

where symbol $\lfloor \cdot \rceil$ denotes a rounding function and parameter s denotes a scaling factor used to enlarge $G(\cdot)$ such that it is an integer. Secondly, Bob convolves the outsourced image with the DoG filters in encrypted domain, the resultant encrypted image can be computed by

$$
\begin{aligned}
G\_[\![f(x, y, \sigma)]\!] &= Enc(G_D(x, y, \sigma) * f(x, y), r_{x,y}) \\
&= Enc(\sum_{u,v} (G_D(u, v, \sigma) f(x - u, y - v), r_{x,y}) \quad \forall u \text{ and } v \\
&= \prod_{u,v} Enc(f(x - u, y - v), r)^{G_D(u,v,\sigma)},
\end{aligned}
$$
$$(6.7)$$

where $r_{x,y}$ is chosen uniformly random and depends upon the location of a pixel on DoG image. Based on Eq. (6.7), we can know that outsourced image $[\![f(x, y, \sigma)]\!]$ convolved with the DoG filters is also the encrypted difference between two Gaussian-blurred images at two neighboring scales. According to the selectively homomorphic cryptosystem, the scaling factor s bounded by its plaintext domain. Thirdly, we need to detect local extrema over encrypted DoG image. Given $D\_[\![f(x, y, \sigma)]\!]$ and its 26 neighbors, this goal can be performed by the *Secure Comparison Protocol* (see Table 6.2).

☐ Generally, SIFT feature descriptor is a 128-dimensional vector. Due to high computational complexity and impractical in the encrypted domain, we use a modify descriptor to generate the weighted magnitudes located at four directions ($0°$, $45°$, $90°$, and $135°$), which constitute a new 4-dimensional vector. For each

4×4 block, let $V(i), i = 0, 1, 2, 3$ denote a feature descriptor vector of 16×16 and $Enc(V(i), r_i) = [\![V(i)]\!]$:

$$[\![V(0)]\!] = [\![V(0)]\!] \odot [\![L_{x_{0°}}]\!] = [\![V(0)]\!] \odot [\![L(x+1, y, \sigma)]\!] \odot [\![L(x-1, y, \sigma)]\!]^{-1}$$
$$[\![V(1)]\!] = [\![V(1)]\!] \odot [\![L_{y_{45°}}]\!] = [\![V(1)]\!] \odot [\![L(x, y+1, \sigma)]\!] \odot [\![L(x, y-1, \sigma)]\!]^{-1}$$
$$[\![V(2)]\!] = [\![V(2)]\!] \odot [\![L_{x_{90°}}]\!] = [\![V(2)]\!] \odot [\![L(x-1, y-1, \sigma)]\!] \odot [\![L(x+1, y+1, \sigma)]\!]^{-1}$$
$$[\![V(3)]\!] = [\![V(3)]\!] \odot [\![L_{y_{135°}}]\!] = [\![V(3)]\!] \odot [\![L(x+1, y-1, \sigma)]\!] \odot [\![L(x-1, y+1, \sigma)]\!]^{-1}$$
$$(6.8)$$

Hence, the keypoint descriptor is $[\![V_k]\!] = ([\![V_k]\!], [\![V(0)]\!], [\![V(1)]\!], [\![V(2)]\!], [\![V(3)]\!])$. To compute $V()$, it's needs *Secure Comparison Protocol* (Table 6.2) to solve.

☐ To match the keypoint descriptor between query image and database over encrypted data, we need to define a distance to introduce the similarity. To conduct this problem, we adopt L_1-distance metric. The L_1 distance between two descriptors V_k^i and V_k^i can be computed by

$$D(V_k^j, V_k^i) = \|V_k^j - V_k^i\|_{L_1} = \sum_{i=0}^{63} |V_k^j - V_k^i| \qquad (6.9)$$

Then, the L_1-distance of this tow descriptors V_k^i and V_k^i over decrypted domain is

$$D([\![V_k^i]\!], [\![V_k^j]\!]) = \|[\![V_k^i]\!] - [\![V_k^j]\!]\|_{L_1}$$
$$= \sum_{k=0}^{63} |[\![V_k^i]\!] - [\![V_k^j]\!]|$$
$$= \prod_{k \in \{t | V_k^i(t) > V_k^j(t), 0 \le t \le 63\}} [\![V_k^i]\!] \odot [\![V_k^j]\!]^{-1} \qquad (6.10)$$
$$\odot \prod_{k \in \{t | V_k^i(t) \le V_k^j(t), 0 \le t \le 63\}} [\![V_k^i]\!] \odot [\![V_k^j]\!]^{-1}.$$

We can use the *Secure Comparison Protocol* (Table 6.2) to compute the above Eq. (6.10). In some works, researchers use Euclidean Distance, Hamming Distance or Similarity to keep the match accuracy. If we use Euclidean Distance D_j in Eq. (6.4) to calculate the matching find, we see strong similarities.

6.3.3 Privacy-Preserving Classification

Classifier learning is one of the most fundamental tasks in machine learning. For instances, in the healthcare system, if we have a rule fits the patient's historical data and if the new patient's data are similar to the historical data, then we can take correct predictions for the new patient's diagnosis. However, the information

Fig. 6.4 Privacy-preserving classification

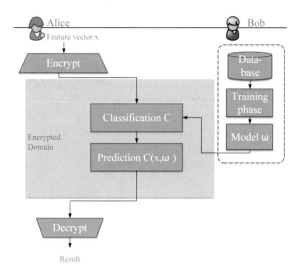

of (new) patient's data are sensitive, including age, gender, past medical history, and current symptoms. Hence, privacy is important if the unlabeled data reflect new patient's diagnosis. To preserve the confidentiality of (new) patients' medical data, a privacy-preserving medical diagnosis should be proposed and the learners maybe pay more attention to guard the crucial privacy of the patients in classification.

Generally speaking, classification model can be broken down into two stages, the *inference stage* in which the server use private training set to learn a private model ω for predicting, and the *decision stage* in which a user want to his private feature vector x can be assigned in optimal class. The model ω is obtained is independent of the user's private input. For example, in Fig. 6.4, Bob (sever) holds database and model ω, user Alice has a feature vector x and prediction result. In this classification model, Bob computes the model ω after running the *training phase* on plaintext data. Only the *classification* needs to be privacy-preserving: Alice should learn $C(x; \omega)$ but nothing else about the model ω, while Bob should not learn anything about Alice's input or the classification result.

For NB classifier, set C contains K classes $\{C_1, C_2, \cdots, C_K\}$, for given a sample x, the conditional risk about x can be computed by

$$g_i(x) = -\mathcal{R}(v_i|x) = \sum_{k=1}^{K} \lambda_{ik} Pr(C_k|x), \qquad (6.11)$$

where v_i denotes the decision to assign the input to class C_i, λ_{ik} denotes the loss incurred for tacking v_i when the input belongs to class C_k, and $\mathcal{R}$ is the expected risk for tacking v_i. The key work is to find a discriminant rule, i.e., using a maximum a posteriori decision rule, works by choosing the class with the highest posterior probability:

$$h^*(x) = \underset{C \in \mathcal{C}}{\mathrm{argmax}}\ Pr(C|x). \qquad (6.12)$$

where $Pr(C|x)$ in Eq. (6.12) can be computed by $Pr(C|x) = \frac{Pr(C)Pr(x|C)}{Pr(x)}$.

Nothing that λ_{ik} and probability $Pr()$ is a real number. Once the classifier model ω is trained by the Bob, the *prediction* program should be operated over encrypted domain to protect the privacy. For private data, Alice multiply a scale factor such that the product are integers before encrypting. Therefore, Eq. (6.12) is operated over encrypted domain.

Right now, Bob has K values $a_1, a_2, \cdots, a_K$ encrypted under Alice' public key and wants Alice knows the argmax over these values, but neither party should learn anything else, where a_i $(i = 1, 2, \cdots, K)$ is integer representation of the $Pr(C_i|x)$ multiply by a scale factor.

Reference

Graepel T, Lauter K, Naehrig M (2012) Ml confidential: machine learning on encrypted data. In: International conference on information security and cryptology. Springer, pp 1–21

Chapter 7
Threats in Open Environment

In the previous chapters, we have introduced solutions of protecting privacy against eavesdroppers in machine learning tasks. In this chapter, we will describe attack techniques of other three types of adversaries.

7.1 Data Reconstruction Attack

Consider the following scenario as federated learning (Fig. 4.1 in Chap. 4), in which each participant holds a local dataset of private information on his own device and tries to build a global machine learning model cooperatively. Each participant adopts the local learning algorithm to trains a model locally and shares only a fraction of model parameters with others. A server acting as an aggregator collects and exchanges these parameters to update the global model at each epoch. Finally, it can create a well-trained model that is almost as accurate as a model built over the union of all the datasets. This approach is considered more privacy-friendly since datasets are not exposed directly.

Hitaj et al. (2017) propose an attack approach against this collaborative deep learning using Generative Adversarial Networks (GANs). This approach allows any trainer acting as a participant to infer sensitive information from a targeted trainer's dataset. In more details, the adversary runs a part of learning algorithm and iteratively reconstructs a subset of data owned by the targeted trainer. Therefore, it can also influence the global learning process and make the reconstructed data more accurate.

The GAN Goodfellow et al. (2014) procedure trains a discriminative neural network D against a generative neural network G. In the original paper of GAN, the discriminative network is trained to distinguish between images of a real dataset and those generated by the generation network. In the initialization, the generative network is randomized by some noises. At each iteration, it is trained to mimic the images that are used for training the discriminative network. Training the GAN is to solve the bilevel optimization problem as follows:

J. Li et al., *Privacy-Preserving Machine Learning*, SpringerBriefs on Cyber Security Systems and Networks, https://doi.org/10.1007/978-981-16-9139-3_7

$$\min_{\theta_G} \max_{\theta_D} \sum_{i=1}^{n_+} \log f(x_i; \theta_D) + \sum_{j=1}^{n_-} \log(1 - f(g(z_j; \theta_G); \theta_D))$$

where x_i are real images and z_j are randomly generated images. Let $f(x; \theta_D)$ be the discriminative network which takes a given image x and parameters θ_D as input and outputs a class label. Let $g(z; \theta_G)$ be a generative network which takes a random noise z as input and outputs an image distribution.

The training procedure is implemented by a local adversarial game. This procedure ends when the discriminative network is unable to distinguish between images of the real dataset and the images generated by the generative network.

7.1.1   Threat Model

The adversary here is an active one that tries to reveal information about a subset of data not belongs to him. To initiate the attack, the adversary pretends to be an honest participant in the collaborative learning task. Different from an eavesdropper, the active adversary can stealthily change the learning process to induce other participant to leak features about the targeted subset.

Hitaj et al. (2017) describe a threat model in detail as follows.

- The adversary is an inside attacker which participates in the privacy-preserving multi-party learning protocol.
- The adversary tries to obtain the information of a subset in which data are labelled with a targeted class, but he does not own this subset of data.
- The adversary does not need to compromise the aggregator that collects parameters and updates the global model. This is a common setting of real-world applications, in which the adversary does not have to control a large set of participants even the aggregator.
- The adversary is active since he does not have to follow the learning protocol benignly. Instead, according to his attack objective, he can directly manipulates values which are uploaded in each epoch. Although the values are not carried out from the local learning process, in the view of other participant, the adversary still uploads and downloads the correct amount of parameters by turn as agreed in advance as required by the collaborative learning protocol.
- In the view of the adversary, the machine learning model is seen as a "white box". That is, he has knowledge of the model structure, parameters, and the range of data labels of other participants.
- The adversary is adaptive to preform a real-time attack during the whole learning task. This attack will last for several rounds. The adversary will be able to iteratively influence the global model by sharing delicate parameters and trick participants into leaking more information on their local data.

7.1.2 Attack Method

Assume that the adversary A acts as a participate of the collaborative learning protocol. According to the protocol, all participants have reached an agreement on a common learning objective. That is, they have known the type of neural network architecture and the range of labels.

The following example considers the adversary and another participant, which can be extend to multiple participants easily. Let P be the target participant that declares that he holds data with labels a and b. Alternatively, the adversary A declares that he has data with labels b and c although A could have no data of the class c. Theoretically, each participant can declare any number of classes, and there is no need for the classes to overlap. The adversary's objective is to reveal as much useful information as possible about the data labelled with a.

During the learning task, the adversary A adopts a GAN to generate data instances which is the forgery of the class a, and pretends to a benign participant. These forged instances are labelled as the class c and injected into the local learning procedure. Other participants can hardly become aware of the missing the class c and thus will "help" the adversary A finish the following learning with the class a and the faked instances. As a result, the global model can gradually recognize whether a given instance belongs to the class a. Thus, the adversary tries to mimic data instance from the class a and uses the current global model to improve his forgery ability about the class a before training. As described in Goodfellow et al. (2014), GANs were initially proposed for density estimation, such that the adversary A can learn the distribution of the data with the class a from the output of the global model without observing the real data directly. In this case, the adversary A construct a GAN according to the global model to reveal more information about the class a which does not belong to him. The details of the attack are shown as follows.

- In this example, there are two participants: an adversary A and a targeted benign participant P. They have agreed on the learning goal and the common structure of the global neural network model $\mathbf{W}$.
- P declares labels a and b, while A declares b and c.
- The collaborative learning protocol will run for several epochs until the global model and both local models achieve an accuracy higher than a certain threshold.
- At the beginning of the benign participant P's training phase, P downloads a percentage of parameters and uses them to update the local model $\mathbf{W}_P$. Then, P trains the local model $\mathbf{W}_P$ with the dataset contains the data of classes a and b. After the local training, P uploads a selection of the parameters of $\mathbf{W}_P$ to the aggregator. The procedure is depicted in Fig. 7.1.
- The adversary A builds a local GAN to mimic the participant P's data of the class a, which is unknown to the from P. At the beginning of A's training phase, A downloads a percentage of parameters and uses them to update both the local model $\mathbf{W}_A$ and the discriminator D. A trains the GAN and generates data instances which is claimed as data of the class c. Then, A trains the local model with the dataset contains data of classes b and (forged) c. After the local training, A uploads

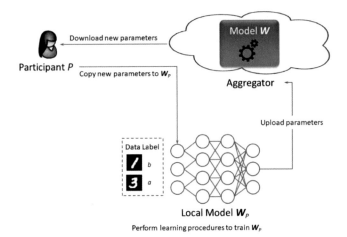

Fig. 7.1 Benign participant's training process

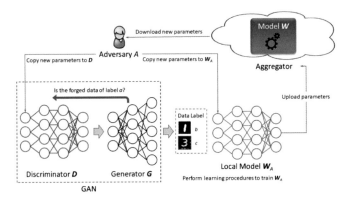

Fig. 7.2 Adversary's training process

a selection of the parameters of $\mathbf{W}_A$ to the aggregator. The procedure is depicted in Fig. 7.2.

The adversary A deals with the extra work of training the local GAN and generating forged data instances. It represents that A tries to learn as much as possible data of the targeted class a.

This attack works as long as the global model and both local models have an improvement on accuracy over time. Some obfuscation techniques such as differential privacy mechanism have a limited effect on defending against this attack, unless they set stronger privacy guarantees, releasing fewer parameters, or establishing tighter thresholds.

So far, this attack can hardly be prevented in the real-time collaborative learning task. However, the attack is usually in an injection way, such that we can detect the attack by testing the final global model if we have the real data of the class c. Although Hitaj et al. (2017) give another approach without the injection, the experimental result shows that it performs worse than the approach that injects forged data in real time.

7.2 Membership Inference Attack

As discussed in Chap. 1, the fundamental question of the membership inference attack is to determine whether a given data record in the training dataset of a specific machine learning model or not. Shokri et al. (2017) consider the machine-learning-as-a-service (MLaaS) setting in which the adversary's access to the machine learning model is in the black-box manner and thus he can only interact with the model by querying with an input and receiving the output. The adversary wants to utilize the information leakage through the prediction output of the model to achieve the attack.

To implement the attack, Shokri et al. (2017) turn the membership inference problem into a classification problem. In another word, the adversary needs to put efforts on training an attack model whose purpose is to distinguish the target model's behaviour on the training inputs from its behaviour on the inputs that are not used for training. Therefore, the adversary can get the inference result by simply inputting the data record into such a model. This attack approach does not concern how the target model was trained.

7.2.1 Threat Model

As the common setting in this book, the machine learning algorithm here is used to train a model that captures distribution of training data. The distribution is represented by the relationship between the content of data records and their labels. These labelled data records in the training dataset are sampled from some population and partitioned into classes.

Given an data record, the well-trained model produces the prediction in the soft-max form. That is, it outputs the prediction vector in which each component shows the probability that the given record belongs to the corresponding class. These probabilities can be seen as confidence values and the class with the highest confidence value is the predicted label of the data record. We usually evaluate the accuracy of the model by testing how it generalizes and predicts data records from the same population but out of the training dataset.

The adversary is allowed to query access to the model and get the model's output (i.e., the prediction vector) on any queried data record. Of course, the adversary knows the format of input records and output predictions, including the domain of input values and the range of output values. The adversary can only have black-box access to the MLaaS platform. That is, he does not know the structure and hyper-parameters of the model and the learning algorithm.

Since the population cannot be hidden, the attacker may have some background knowledge about the population of the training dataset of the target model. That means he can sample data from the population and thus mimic the training dataset in some degree. That also means the attacker can learn some general statistical property of the population.

At the beginning of the attack, the adversary is given a data record and the black-box access to the target model. The adversary's goal is to correctly determine whether this record is a member of the training dataset of the model or not.

7.2.2   Attack Method

Generally, the overfitting is inevitable for a machine learning model, such that the model will give different outputs when inputting data in the training dataset and data not in the dataset. It is intuitive that the adversary may construct an attack model to correctly perform member inferences on the training dataset by utilizing this property.

To train the attack model, the adversary should build several models each of which is called shadow model and has a similar behaviours to the target model. Since these shadow model are built by the adversary, he naturally knows whether a given record was in any shadow model's training dataset or not. It is a big difference between the target model and shadow models. As a result, the adversary can see each input and its corresponding output as an instance (each labelled "in" or "out") and perform a supervised learning over a set of instances. After training, the adversary can obtain the attack model which can distinguish outputs on members of shadow models' training datasets from outputs on non-members of the datasets.

Let function $f_{target}()$ be the target model, and let D_{target}^{train} be its training dataset which contains several labelled data records $\{(\mathbf{x}^i, y^i)\}_{target}$. The target model takes a data record $\mathbf{x}^i_{target}$ (its true label is y^i_{target}) as input and outputs a probability vector of size c_{target}. The components of this vector are in the soft-max form, i.e., in $[0, 1]$ and sum up to 1.

Once the attack model is built, the adversary can initiate the attack as depicted in Fig. 7.3. Given a labelled record $(\mathbf{x}, y)$, the target model can produce the prediction vector $\mathbf{y} = f_{target}(\mathbf{x})$. The distribution of confidence values depends on the true label of $\mathbf{x}$ in some degree. Besides the prediction vector $\mathbf{y}$, the true label y is also regarded as an input of the attack model.

Given probabilities $\mathbf{y}$ and the label y, the attack model outputs the probability $Pr((\mathbf{x}, y) \in D_{target}^{train}$, i.e., $\mathbf{x}$ is in the training dataset of $f_{target}()$.

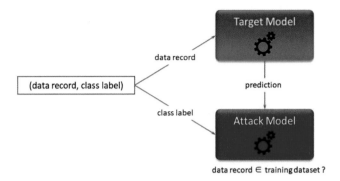

Fig. 7.3 Membership inference attack in the black-box setting

7.2.2.1 Model Building

To train shadow models, the adversary has to obtain a set of training data which have the similar distribution to the training data of the target model. Shokri et al. (2017) give some methods for generating such data.

Model-based synthesis. The adversary can generate synthetic training data for the shadow models using the target model itself. Each training instance is comprised of a chosen data record $\mathbf{x}$ and the label y in accordance with the probability vector $\mathbf{y} = f_{target}(\mathbf{x})$. The intuition is that records that are classified by the target model should be similar to that in the training dataset. Therefore, they are suitable for feeding shadow models.

Statistics-based synthesis. The adversary may have some statistical information about the population of training data of the target model. With this condition, the adversary can generate synthetic records by independently sampling the record's value of each feature from its own marginal distribution. The shadow models trained on these records make the attack model more effective.

Then, the adversary tries to build k shadow models $\{f_{shadow}^i()\}$. The i-th shadow model is trained on the i-th dataset D_{shadow}^i generated by one of the methods above. The worst case for the adversary is that synthesis training datasets of shadow models are disjoint from the training dataset of the target model, i.e., $D_{shadow}^i \cap D_{target}^{train} = \emptyset$.

Normally, shadow models should be trained in a similar way to the target model. If the adversary knows the learning algorithm and the structure of the target model, the training task will be easier. Although the information above is unknown in the MLaaS, the shadow models can still be trained by using the same platform (e.g., Machine Learning API in Fig. 7.4) as was used to train the target model. The parameters of all models are trained independently. It is intuitive that the more shadow models, the more accurate the attack model will be. Therefore, the adversary should be careful to set the value of k for achieving an acceptable attack effect.

Shokri et al. (2017) present that the intuition of the shadow model technique is that similar models trained on similar data records using the same learning service in a

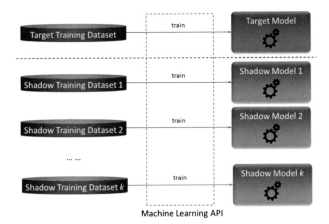

Fig. 7.4   Training shadow models

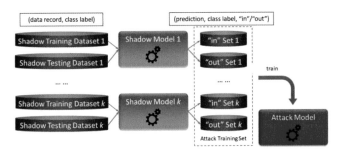

Fig. 7.5   Training attack models

similar way. This observation is empirically demonstrated out by their experiments. The next task of the adversary is to build a model that can infer membership in shadow models' training datasets since he has owned all these training datasets.

For each shadow model, the adversary sets two disjoint datasets. The one is its own training dataset and the other is the testing dataset of the same size. The adversary queries the model with the two datasets and obtains the outputs. The outputs on the training dataset are labelled with "in", while that on the testing dataset are labelled with "out". After labelling, the adversary has a dataset of instances each of which contains a data record, the corresponding output (i.e., probability vector) of a shadow model, and an "in/out" label. The attack model takes the record and the corresponding output as input and outputs an inference of the "in/out" label.

The procedure of training the attack model is depicted in Fig. 7.5. For generating training instances for the attack model, the adversary queries shadow models. In more details, for each i-th shadow model and each $(\mathbf{x}, y) \in D^i_{shadow}$, the adversary computes the prediction vector $\mathbf{y} = f^i_{shadow}(\mathbf{x})$ and insert the instance $(y, \mathbf{y}, \mathbf{in})$ into the attack model's training dataset D^{train}_{attack}. Let T^i_{shadow} be the i-th testing dataset disjoint from the i-th training set D_{shadow}^i. For each i-th shadow model and each $(\mathbf{x}, y) \in T_{shadow}^i$, the adversary computes the prediction vector $\mathbf{y} = f_{shadow}^i(\mathbf{x})$ and insert the instance $(y, \mathbf{y}, \textbf{out})$ into the attack model's training dataset D_{attack}. Then, the adversary splits the dataset D_{attack} into c_{target} partitions each of which corresponds to a different class label. Thus, for each label y, the adversary trains an attack model which takes $\mathbf{y}$ as input and predicts the membership of $\mathbf{x}$ by outputting "in" or "out".

Although the attack model is trained on the data with high confidence, it does not simply learn to recognize inputs that are classified with high confidence. Actually, the attack model is to distinguish between the training inputs and non-training inputs, where all the inputs are classified with high confidence.

Indeed, from the functionality of the attack model, the membership inference problem is converted into a binary classification problem. The binary classification "Yes or No" is a standard machine learning task, and there is many way for build the attack model, e.g., using the state-of-the-art machine learning framework. The adversary has no need to specify methods used for training the attack model, but can still obtain a effective model. Moreover, he has no control over the model structure, model parameters, or training hyper-parameters. Evaluating the attack accuracy is to test the precision (what fraction of records inferred as members are indeed members of the training dataset) and the recall (what fraction of the training datasets' members are correctly inferred as members by the adversary).

Shokri et al. (2017) also point out that the model achieving $\epsilon-$ differential privacy are secure against membership inference attacks, since such attacks make decisions on the outputs of the model rather than auxiliary information of data records. However, differentially private models may significantly reduce the model's prediction accuracy for small ϵ values.

7.3 Model Stealing Attack

Besides building a machine learning model on the training dataset, the MLaaS also allows model owners to charge others for queries to their commercially valuable models. Therefore, the contents of such models are often coveted by curious adversaries. Tramèr et al. (2016) explore an attack named model stealing attack, in which the adversary tries to break the confidentiality of machine learning models through query access. In another word, the adversary's goal is to extract an equivalent or near-equivalent model of the target model in the functionality.

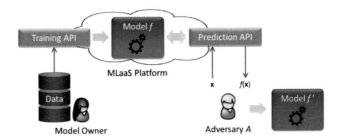

Fig. 7.6 Model stealing attack

7.3.1 Threat Model

In the model stealing attack (Fig. 7.6), the adversary attempts to learn a target model f ideally and has black-box access to the model f. Actually, obtaining a model f' that closely approximates f is enough.

The adversary has some motivations to perform the model stealing attack, such as avoiding query charges in MLaaS, violating the privacy of training data, and facilitating evasion attacks in the detection.

In practice, there are two modes for the query of the adversary. The one is making direct queries with an arbitrary data record $\mathbf{x}$ and obtaining the model's output $f(\mathbf{x})$. The other is making only indirect queries with points in the input space and getting outputs of a certain distribution. In this mode, the adversary may not know the feature extraction mechanism of meta data records. These two modes of queries arise in the MLaaS. The output of the model f value can be a probability vector in the soft-max form or a class label, which depends on the public machine learning API of the MLaaS.

Similar to some security schemes, the adversary A here can be modelled as a probabilistic algorithm. The adversary's objective is to efficiently carry out an approximation model f' that closely matches the target model f by querying f as few as possible.

Tramèr et al. (2016) give two error measures to formalize "closely matching".

- **Test error R_{test}.** This metric represents the average error over a test set D. The test error between f and f' is $R_{test}(f, f') = \sum_{(\mathbf{x}, y) \in D} d(f(\mathbf{x}), f'(\mathbf{x}))/|D|$. The lower the test error is, the better the matching for D (distributed like training dataset) between f' and f is.
- **Uniform error R_{unif}.** This metric represents the average error over a set U where each record is uniformly chosen in its feature space. The uniform error between f and f' is $R_{unif}(f, f') = \sum_{\mathbf{x} \in U} d(f(\mathbf{x}), f'(\mathbf{x}))/|U|$. The error R_{unif} estimates the fraction of the full feature space on differences between f' and f.

Therefore, the attack accuracy in two measures can be define as $1 - R_{test}(f, f'$ and $1 - R_{unif}(f, f'$, respectively. Obviously, the accuracy here is related to the distance function d which shows how close the class probabilities output by f' are to those of f.

The adversary A is assumed to have no more information about the target model's training data other than that provided by the machine learning API of the MLaaS platform. Thus, if A believes that f belongs to some model class, then A intuitively constructs the model f' from the same class.

7.3.2   Attack Method

Tramèr et al. (2016) present two kinds of model stealing attacks by focusing on the prediction vector output by f. This output of a query $\mathbf{x}$ to f falls in a range $[0, 1]^c$ where c is the number of components each of which indicates the probability that $\mathbf{x}$ belongs to the corresponding class.

Equation-solving attacks. Normally, many machine learning models directly carry out class probabilities as a multi-value function of the input $\mathbf{x}$. Such a function works with wreal-valued parameters. The access to f provides these class probabilities to the adversary A with several synthesis data in the form as $(\mathbf{x}, f(\mathbf{x}))$, by which A can build equations with unknown model parameters. For some models, these equation systems can be efficiently solved, and thus the approximation f' can be constructed by the parameters. The attack result can only be evaluated by the experiments primarily. Sometimes, the input should be pre-processed by removing the rows with missing values, encoding categorical features into the one-hot form, or scaling numeric features into $[-1, 1]$.

Decision Tree Path-Finding Attacks. For the decision tree, the output is not a vector of class probabilities. Different from a continuous function, a decision trees partitions the input space into discrete regions, each of which is assigned a class label and a confidence score. That is each leaf of the tree is labelled with a class label (a real-valued output for regression trees) and a confidence score. The adversary A produces a pseudo-identifier for the path that the input traversed in the tree through the query access to the target model f. By varying the value of each input feature, he then finds the satisfied predicates to make an input follow a given path in the tree. Considering there may be some in-completed inputs with missing features, A can label each node in the tree with an output value which is used as the output if the tree traverse of an in-completed input terminates in this node.

References

Goodfellow I, Pouget-Abadie J, Mirza M, Xu B, Warde-Farley D, Ozair S, Courville A, Bengio Y (2014) Generative adversarial nets. In: Advances in neural information processing systems 27

Hitaj B, Ateniese G, Perez-Cruz F (2017) Deep models under the gan: information leakage from collaborative deep learning. In: Proceedings of the 2017 ACM SIGSAC conference on computer and communications security, pp 603–618

Shokri R, Stronati M, Song C, Shmatikov V (2017) Membership inference attacks against machine learning models. In: 2017 IEEE symposium on security and privacy (SP), IEEE, pp 3–18

Tramèr F, Zhang F, Juels A, Reiter MK, Ristenpart T (2016) Stealing machine learning models via prediction apis. In: 25th {USENIX} Security Symposium ({USENIX} Security 16), pp 601–618

Chapter 8
Conclusion

So far, the privacy-preserving machine learning has been a fruitful and long-standing research topic in both academic community and industry. With the development of cloud computing and the increasing concern of data privacy, we believe that more and more researchers will focus on this hot topic. In this book, we present some typical researches on privacy-preserving machine learning techniques, while it is far from complete due to space and time constraints. In the following, we present several possible open problems to stimulate further research.

- **Flexible Encryption Design**. As we discuss in previous chapter, there have been a large number of encryption schemes for the secure computation in the open environment. Currently available solutions mostly focus on one or two encryption approaches for a given learning task, while mainly emphasizing the efficiency. Is it possible to find an efficient algorithm for cryptographic operations generally or flexibly used in machine learning tasks? Is the acceleration method of the non-privacy-preserving machine learning task also suitable for the privacy-preserving one?
- **Massive Data**. In some of learning scenarios, the original data from different resources (data owners) are not correlated. Hence, they should be pre-precessed before the a collaborative learning task. This topic is rarely discussed in existing literature. How can we achieve some pre-precession operations (e.g., alignment and intersection) on such massive data while protecting the privacy of each individual data owner?
- **Dynamic Learning**. The collection of data in a collaborative learning task is sometimes dynamic, so as to the participants. Besides protecting the data privacy, there is a need to be able to handle quickly changing data and the unstable group of participants. Here, the data security heavily depends on the reliability of the learning system. How to deal with the real-time events, such as the drop of participants, is another open problem for privacy-preserving learning tasks.

© The Author(s), under exclusive license to Springer Nature Singapore Pte Ltd. 2022   87
J. Li et al., *Privacy-Preserving Machine Learning*, SpringerBriefs on Cyber Security Systems and Networks, https://doi.org/10.1007/978-981-16-9139-3_8

- **Misbehavior Detecting**. In Chap. 7, we can learn that most attacks are caused by adversaries who are active to violate the learning protocol. How do we detect the misbehavior of any untrusted participant in the large-scale learning task? Indeed, we can make a test on the result by using our experiences to check its validity after the learning task is finished. However, in the worst case, it is meaningless at this time. Therefore, the real-time detection is still an open problem.

Printed in the United States
by Baker & Taylor Publisher Services